THE AMAZON RIVER FOREST

THE AMAZON RIVER FOREST

A NATURAL HISTORY OF PLANTS, ANIMALS, AND PEOPLE

Nigel J. H. Smith

New York Oxford

Oxford University Press

1999

Oxford University Press

Oxford New York
Athens Auckland Bangkok Bogotá Buenos Aires Calcutta
Cape Town Chennai Dar es Salaam Delhi Florence Hong Kong Istanbul
Karachi Kuala Lumpur Madrid Melbourne Mexico City Mumbai
Nairobi Paris São Paulo Singapore Taipei Tokyo Toronto Warsaw

and associated companies in
Berlin Ibadan

Library of Congress Cataloging-in-Publication Data
Smith, Nigel J. H., 1949–
The Amazon River forest : a natural history of plants, animals,
and people / Nigel J. H. Smith.
p. cm.
Includes bibliographical references and index.
ISBN 0-19-512207-0 — ISBN 0-19-512683-1 (pbk.)
1. Land use—Amazon River Region. 2. Human ecology—Amazon River
Region. 3. Natural history—Amazon River Region. 4. Land use—
Brazil, North. 5. Human ecology—Brazil, North. 6. Natural
history—Brazil, North. I. Title.
HD469.A43S6 1999
333.73'13'09811—dc21 98-4759

9 8 7 6 5 4 3

Printed in the United States of America
on acid-free paper

To the memory of
James Jerome Parsons
(*1915–1997*),
geographer, historian, teacher

Preface

The approach adopted in this book combines the strengths of natural history studies with attention to more "modern" concerns such as land-use dynamics, forces driving landscape change, and policy implications of social and ecological change. Although some people regard natural history as a bit old-fashioned, the focus here is on the natural history of *people*. It is increasingly difficult to find accounts of development and socioeconomic change in Amazonia that provide much of a feel for how people actually live and eke out their livings in the diverse landscapes of the region; scholars often shy away from such "descriptive" endeavors. But understanding the rich tapestry of cultures in Amazonia is essential if we are to devise sound development projects and policies. This book incorporates historical influences, the impact of rapid urbanization, and the introduction of new technologies into its exploration of Amazonian lifeways and the natural resource base that supports them.

A map of the Amazon River area can be found as figure 1.1. Because I occasionally mention various states of Brazil, both within Amazonia and in the northeast and southern regions, I have also provided a map of the states of Brazil (figure 2.1). In an attempt to ease the flow for readers, the text of this book generally uses only the common names of Amazonian plants and animals. Scientific names of plants and animals are relegated to appendices A and B, where they may provide useful checklists for fieldworkers on the Amazon floodplain in Brazil. (I should point out, however, that the taxonomy of many groups of animals and plants needs further work, and that the scientific names should therefore be regarded as provisional, or "best bets" only.) When two or more plant species share the same common name, such as is the case with tucumã palm or bacuri fruit, I provide the scientific name in the text to avoid confusion. "Further Reading," at the end of the book, groups bibliographic sources by chapter and by brief descriptive headings that correspond to different subject matters within the chapters.

I was first introduced to the Amazon floodplain in 1970 during a field course to the Brazilian Amazon led by Hilgard O'Reilly Sternberg, now professor emeritus at the University of California, Berkeley. Professor Sternberg did pioneering work on the settlement and landforms of Careiro Island near Manaus, and I was privileged as an undergraduate student to witness the lifeways of rural folk on the Amazon floodplain near Manaus and Santarém. Although most of my own fieldwork in the early 1970s involved upland colonization schemes in the Amazon, I made a point of periodically visiting Careiro and its environs on my journeys in and out of Brazil.

The late Professor James Parsons of the Department of Geography at Berkeley, to whose memory I dedicate this book, provided inspiration and encouragement for me to study the natural history of people in Amazonia. Charles Bennett and Jonathan Sauer, both emeritus professors of the Department of Geography at the University of California, Los Angeles, and Professor Hartmut Walter, also of the Department of Geography at UCLA, were also instrumental in fostering my concern for biodiversity and its uses and abuses by different cultures.

Numerous other individuals and organizations have enabled me to conduct intermittent fieldwork along the Amazon River for over a quarter of a century. In 1976, Warwick Kerr invited me to work at the National Institute of Amazonian Research (INPA; Instituto Nacional de Pesquisas da Amazônia) in Manaus. Under Dr. Kerr's directorship at INPA, I was able to make numerous field trips on the Amazon floodplain in the state of Amazonas between 1976 and 1980. While serving as a researcher at INPA, I lived in Itacoatiara, a small riverside town on the north bank of the Amazon, for most of 1977. Although my studies there focused on subsistence and commercial fishing, I was able to make field observations on land use along a 100-kilometer stretch of the middle Amazon floodplain during the height of the jute boom. I made brief return trips to Itacoatiara in 1978 and 1979, as well as short field excursions on the Amazon floodplain between Manaus and Manacapuru in those years.

From 1991 to 1996, I made a dozen boat trips on the floodplain of the middle and lower Amazon to interview farmers and ranchers. The field trips were not prolonged, typically lasting less than 2 weeks, but they spanned all stages of the rising and subsiding of water levels, thereby providing me with opportunities to observe various phases of the agricultural calendar and the seasonal rhythm of forest gathering. I have been on the Amazon during a year of very high floods (1976), a year of rapidly rising floods (1994), shortly after a major flood (1997) and during exceptionally severe dry seasons with low water levels (1992 and 1995). I made the following trips to the Amazon floodplain between 1991 and 1997 (sponsors/funding agencies in parentheses): Breves/Caxiuanã area, November 1995 (Museu Goeldi); Combu Island near Belém, November 1994 and December 1995 (1-day field trips following conferences); Itacoatiara, November 1991 (EMBRAPA—Empresa Brasileira de Pesquisa Agropecuária); Macapá area, December 1994 and May 1996 (conference and Pilot Program to Conserve the Brazilian Rainforest, World Bank); and the Santarém area, September/October 1992, November 1992, March, 1993, August/

September 1993, June 1994, May 1996, July 1996, September 1997 (CIAT—Centro Internacional de Agricultura Tropical, World Bank, Department of Geography, and Center for Latin American Studies at the University of Florida). Zenaldo Coutinho, a representative in the state assembly of Pará, kindly arranged for a state government airplane to take me to Afuá in northwestern Marajó Island in September 1997. The bulk of my field observations in the 1990s were made along a 300-kilometer stretch of the Amazon from Nhamundá to Monte Alegre.

Many individuals helped arrange logistical support and provided advice for my fieldwork. I am particularly grateful to Pedro Marques (SENAR—Serviço Nacional de Aprendizagem Rural) in Santarém for his assistance in procuring various government-owned or privately operated boats. A trained agronomist, Pedro accompanied me on several excursions. Because he grew up on a jute farm on the Amazon floodplain, he was able to provide many useful insights on the agricultural economy of the middle Amazon. Dilson Frazão, Luciano Marques, Gladys Martinez, and Adilson Serrão of EMBRAPA–Amazônia Oriental were also most helpful in arranging logistical support for fieldwork in the Santarém area. Italo Falesi, also of EMBRAPA, provided advice on sites for field visits and arranged for the analysis of soil samples.

Ruy Corrêa, the ex-mayor of Santarém and a rancher on the Amazon floodplain, graciously exchanged views with me on agricultural development along the river. I am most grateful to Delano Riker for arranging for a 3-hour over-flight of the Amazon floodplain between Óbidos and Monte Alegre in May 1996 and for discussing the prospects for fish farming using indigenous fruits. In Oriximiná, José Mileo, a local rancher, provided useful information on the history of cattle and water buffalo ranching in the area of the confluence of the Trombetas and Amazon.

Eloisa Cardoso of EMBRAPA, Belém, and Carlos Iglesias of CIAT in Cali, Colombia, kindly checked the manioc databases of their respective institutions for accessions in their collections with names that matched cultivars I noted in the Brazilian Amazon.

Mark McLean prepared the maps with the assistance of Susan Swales. Susan Swales also made helpful comments on an early version of the manuscript. Joyce Berry, Senior Editor with Oxford University Press, helped mold the manuscript into a more readable form with her creative editing and organizing suggestions. I am indebted to Joyce for going beyond the call of duty in helping to identify many ways to improve the book's flow and coherence. The copyeditor, Beth Macom, also made many helpful suggestions for reorganizing text and overcoming deficiencies in my penmanship.

I do not wish to imply that any of the above organizations or individuals agrees with my findings or endorses this study. The conclusions and interpretations are entirely my own.

Gainesville, Florida N. J. H. S
January 1998

Contents

THE AMAZON RIVER FOREST

1

Biodiversity as a Cornerstone for Agricultural Intensification

With global demand for agricultural products expected to double or even triple within the next fifty years, productivity of existing crops and grazing lands will have to be intensified dramatically. The alternative—agricultural *extensification*, the opening of new lands for production—is not a viable long-term option, because it would eliminate many remaining habitats for wildlife and destroy environments that provide valuable ecological services for human society. At present, both processes are underway throughout the world. In some areas, agricultural production is being intensified; in others, farmers spill into forests, savannas, and wetlands to cut, plow, and dike them for fields and livestock grazing. The productivity of already-cleared areas needs to be improved, rather than promoting the opening of new farmlands.

How the process of agricultural intensification plays out over the next few decades will largely determine how many species of plants and animals—the products of millions of years of evolution—will survive in the twenty-first century. While few people would argue against the need to intensify crop and livestock production systems, the concept of intensification means different things to different constituencies. For some, it implies a greater use of pesticides, herbicides, fertilizers, and other purchased inputs. Others argue that because such an approach exacerbates collateral damage to the environment (for example, the poisoning of water supplies for humans and wildlife), it is thus unsustainable. These advocates of "alternative" or "traditional" agriculture call for an agroecological approach to intensification that emphasizes recycling of nutrients, mulching, and using nonchemical approaches to pest control.

After more than a decade of discussions about sustainable agriculture and development the concept remains nebulous. "Sustainable" agriculture has become a kind of soup du jour; its flavor is highly variable, depending on the agendas of the cooks in charge. We want to be believers, but are not sure what we are supposed to believe in. Rather than attempt to shed new light on the sustainability debate, however, this book adopts a more modest approach. Its main message is that biodiversity is an essential resource for adapting agricultural systems to shifting ecological and socioeconomic conditions, and that local knowledge is an often overlooked resource for the better management and conservation of biological resources.

The path to more enlightened agricultural development lies between the two extremes of heavy reliance on chemical inputs and more traditional approaches to agriculture. A new paradigm for agricultural research and development is emerging that combines the judicious use of chemical inputs with greater reliance on the management of biological resources. Conservation and better management of biodiversity is critical to boost and maintain crop and livestock yields. Furthermore, the safeguarding of habitats for wild plants and animals provides options for the domestication of new crops and livestock to cater to emerging markets, or to adapt agriculture to challenging environments. Multiple cropping, increased genetic variation among crops, integrated pest management, and a mosaic of land uses are among the ingredients of biodiversity-enhancing approaches to intensification.

The Amazon floodplain, or *várzea*, whose rich alluvial soils represent the last major agricultural frontier of the Americas, provides a propitious setting for examining the interplay between biodiversity, agricultural intensification, and indigenous knowledge. Because the Amazon River is culturally and ecologically heterogeneous, any discussion of land use and agricultural intensification needs to consider such variation (figure 1.1).

Agroecology of Landscapes along the Amazon

The headwaters of the 3,000-kilometer-long Amazon begin high in the Andes, where erosion gnaws away at the mountain chain thrust up by the collision of the South American and Pacific plates. This megascale geologic event has major implications for agriculture and the productivity of fisheries along the Amazon. Fertile volcanic soils and marine sediments are scooped from steep mountainsides by torrents cascading down the deep gorges of the Andes. These sediments are deposited and picked up again and again by the sweeping and unpredictable currents of the mighty river that conveys nearly one-fifth of the fresh water flowing off the face of the earth. No river comes close to discharging as much water as the Amazon, particularly during its flood stage, when the Amazon pushes creamed-coffee–colored water hundreds of kilometers out into the blue Atlantic. Not surprisingly, the Amazon floodplain is also the world's most generous. Along some stretches during the flood stage, one can paddle from one side of the floodplain to the other for 50 kilometers without being able to step on firm ground (figure 1.2).

The annual flood along the Amazon is the main environmental pulse that determines the rhythm of cultural activities and life cycles of plants and animals.

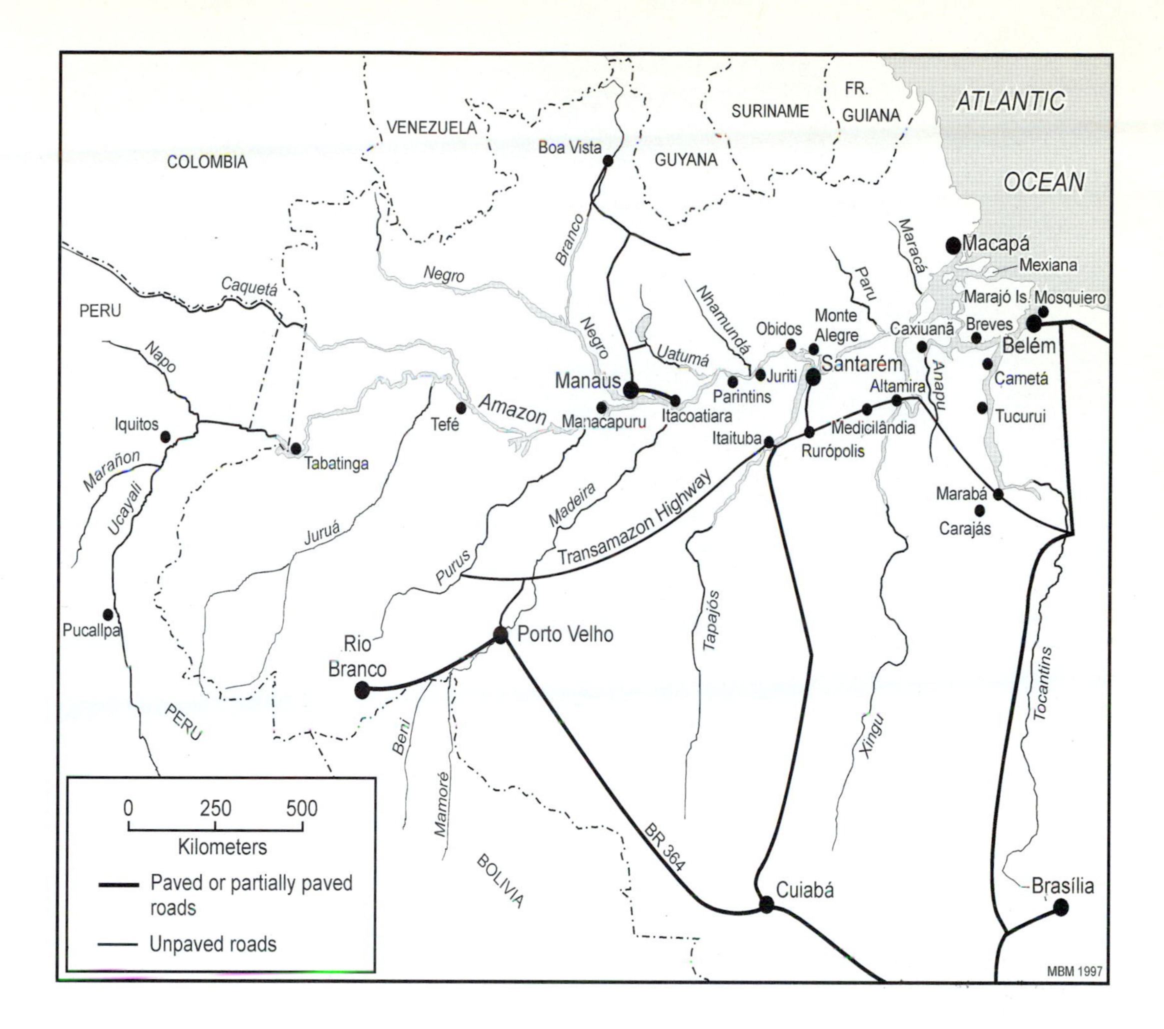

Figure 1.1. The Amazon Basin with some of its principal urban centers and highways.

The yearly surge of floodwaters creates a seasonal river forest (figure 1.3). In backswamps along the margins of lakes, fisherfolk can canoe through the forest canopy for several months, whereas on the higher banks, trees may only be partially submerged, and that for a few weeks only. Along the lower stretches of the Amazon, the seasonal difference between high and low water is only 2–4 meters, whereas along the middle Amazon around Manaus, the annual change in amplitude of water level is around 10 meters. The greatest annual fluctuation in water levels appears to occur around Tefé—some 12–15 meters—and then decreases upstream to around 6–7 meters at Iquitos.

The Amazon is a great sculptor, constantly molding and reshaping its floodplain, creating new islands and mud bars, tailoring others to its changing currents, and forming new lakes while filling in others. But far from negating the possibility of productive agriculture along its floodplain, the Amazon's annual floods are a blessing, bringing fresh deposits of silt and nutrients to soils that were farmed with annual crops during the preceding dry season, and providing perennials with a nutrient recharge. Agricultural activities on the Amazon are concentrated on the highest part of the floodplain: the main banks of the river and its associated side-arms (*paranás*) that snake around an island or cut across lakes.

Figure 1.2. The floodplain at Lago Grande de Monte Alegre, Pará, mostly deforested for cattle ranching, about a month before the crest of the annual flood. Meadows in this area have been burned periodically to promote pasture for at least 150 years. The lake in the background is a commercially important fishery, now threatened by excessive catches and loss of habitat for fruit-eating fish at high water. May 1996.

Figure 1.3. A fisherman alongside the forested banks of the Rio Tapará near Santarém, Pará. The water level has been dropping for about two months, leaving aquatic vegetation draped from the branches of one of the trees. A couple of months later, the man will be able to walk through the forest. A bow, one of the fishing implements on board, is in the middle of the canoe. September 1993.

Figure 1.4. A backswamp forest on the Amazon floodplain, now partially cleared for grazing cattle and water buffalo. In the background is the scarp of the Santarém plateau (*planalto*). Some buriti palms in the center have been spared from the chainsaw because they produce edible fruits, but many other trees with useful products have been lost. The waters of the Amazon are rising and have spilled over into the remnant backswamp forest and pasture. Rio Maicá near Murumuru, Municipality of Santarém, May 1996.

The banks are relatively steep closest to the river, then flatten out and taper off gradually into a lake or backswamp (figure 1.4). The banks of channels (*furos*) that lead from the river or side arm to a lake are not as high but are often cultivated seasonally. At the same time as one bank is suddenly undercut by the temperamental Amazon (figure 1.5), bringing crops and even houses tumbling into the river, a fresh spit or island is forming somewhere else. Farmers take advantage of such "new lands" to grow seasonal crops.

Farmers have thus learned not only to roll with the seasonal variations in water level, but to adapt to constant changes in the lay of the land. Some farmhouses are now on old banks, long since cut off from the main river channel. A washboard of new banks and swales separates the houses from the Amazon, a gift of fine farmland as the Amazon migrates in another direction. Differences in elevation on the floodplain create a mosaic of microenvironments for plants, both wild and cultivated. In one community along the Amazon in Peru, for example, farmers recognize six distinct environments for planting, ranging from the backslopes of levees to sand bars. And in the Amazon estuary, farmers have learned to contend with tides, another leitmotif superimposed on the annual rise and fall of the river. Tidal influences are felt as far upstream as Óbidos in Pará, but are really only a source of contention for farmers on the myriad islands at the broad mouth of the Amazon.

Figure 1.5. Along the Amazon, periodic upwelling currents create smooth lenses on the surface. The billowing currents steepen banks along the river, eventually causing them to collapse. According to legend, some of these currents are caused by the stirrings of an enormous underwater snake called cobra grande. Upwelling currents can be treacherous for canoes and are testament to the power of the river as it constantly reshapes the floodplain. Near Santarém, Pará, March 1993.

The Amazon Floodplain and Agricultural Development

In the 1960s and 1970s, development planners targeted the Amazonian *uplands* as their main arena for lifting the region out of its economic doldrums. This proclivity to emphasize development on the *terra firme* was fueled by a desire to put a stamp of sovereignty on the sparsely settled region. The building of roads was perceived as a "modern" way to open up a region for development. But experience with colonizing and developing the Amazon uplands has brought mixed results. While some farmers and ranchers have prospered, others have not fared so well, and in many cases have invaded parks and reserves in search of profits. As the tempo of forest destruction increased markedly throughout the 1970s and 1980s, it triggered international concern. Efforts to further open up the *terra firme* have consequently slowed.

And so, after more than two decades of upland development, policymakers and investors are increasingly looking to the Amazon floodplain as a new venue for development. As the traditional bias against the *várzea* appears to be changing, the Amazon floodplain is experiencing dramatic ecological and cultural change. Although the Amazon floodplain accounts for a relatively small proportion of the region's land mass, perhaps less than 5 percent, it still embraces nearly 300,000 square kilometers. For decades, a few local politicians and scientists have recognized that the *várzea* is an especially favorable environment for agricultural development because of its rich soils and ease of communication. Yet sporadic development efforts have come and gone, all of them narrowly focused, such as the promotion of rubber collecting in the nineteenth century and jute cultivation in the twentieth. Such spur-of-the-moment efforts at development rarely have any lasting impact, and Brazilians have coined an appropriate term for this impatient approach: *imediatismo*.

In the past, the seasonally flooded Amazon plain has supported dense human populations organized into sophisticated chiefdoms, but at present its agricultural potential is being undercut by a wave of cattle and water buffalo ranching. The process of transforming the landscape of the Amazon floodplain into extensive tracts of grassland for livestock grazing is obliterating species-rich forests and small farms with diverse cropping systems.

Floodplain forests contain many plants that are gathered for a wide variety of uses. Some of these plants are also cultivated on a small scale in home gardens. As the forest shrinks, however, options for diversifying agriculture are curtailed. Because small farms are often rich islands of agrobiodiversity, their loss results in the disappearance of plants in early stages of domestication. Also lost is farmers' traditional knowledge about a wide range of natural resources associated with farming the varied topography of the Amazon floodplain. The Amazon River is becoming a highway to landscape homogenization.

All is not yet lost, however. The successes of some innovative farms and even ranches suggest ways to diversify agricultural and livestock production so that yields increase and incomes rise. True, improved living standards do not guarantee diminished pressure on remaining forests and other habitats important for wildlife, especially fish; successful farmers or ranchers might be tempted, for example, to clear all their properties. But intensification can be compatible with conservation of habitats and their natural resources, and when living standards rise, people have more options. Both farmers and ranchers are now experimenting with novel crops and cropping patterns, as well as game and fish farming. Development planners and conventional agricultural research and extension systems generally overlook these locally conceived and better-adapted approaches to agriculture and livestock production.

Progress in feeding humanity and satisfying growing consumer demand for a wide variety of other agricultural products rests neither in blind faith in science, nor a simple return to traditional farming. What is required to meet the challenge of boosting agricultural productivity along the Amazon and elsewhere is a blend of scientific research and indigenous knowledge.

While the great potential of the Amazon floodplain as a source of staples, fish, and other natural products has long been recognized, no systematic effort has been made to elucidate its diverse ecological and cultural attributes as a foundation for rational development. The lack of appreciation for the cultural and ecological nuances of the Amazon floodplain is no doubt a major reason why so little has been accomplished in tapping its full potential for all residents of the basin.

Without a deeper understanding of the complex interactions of people and their environment along the Amazon, future development projects may promote destruction of the very cultural and ecological resources required for long-term economic growth and conservation. This book makes no attempt to catalogue the myriad ways in which people exploit resources along the Amazon, nor to identify their numerous ecological impacts. Rather, it explores the main land-use systems along with some of their resource management issues. The discussion highlights major environmental repercussions of agricultural activities across the diverse land-use systems, and discusses policy implications for biodiversity conservation and resource management.

Plan of the Book

This book explores the ways in which local populations depend on the remaining tracts of forests and other plant communities along the Amazon. The loss of such biological resources as a result of habitat disturbance, particularly in service of cattle and water buffalo ranching, means far more than an abstract loss of biodiversity. Of most immediate impact is the fact that people are forced to purchase fruits, construction materials, and plant remedies from other sources when local supplies are eliminated. But the survival of forests and other habitats is not only important as a source of dietary enrichment, household supplies, and income generation; it is vital for sustaining agricultural productivity over the long term. Because floodplain forests contain wild populations of some existing crops, they act as reservoirs for potentially useful genes for crop improvement. Further, the forests support the recruitment of new crops as men, women, and children deliberately transplant forest seedlings into their home gardens. Chapter 1 has provided an introduction to the challenge of preserving plant communities while achieving agricultural intensification along the Amazon.

Chapter 2 examines the history of occupation of the Amazon River and its adjacent uplands from earliest times through the modern period. This historical backdrop to settlement in the region provides some valuable lessons on carrying capacity and natural resource management. The main points emerging from this discussion are that the Amazon River has been settled for a long time, that human populations have often been dense, and that little if any of the forest along the river and adjacent upland is pristine. At the time of first contact with Europeans, the Amazon floodplain and surrounding *terra firme* were much more densely settled than at present, but much more forest—at various stages of succession following clearing—was left standing. Over time, however, much of the tree cover was cleared and the fallow vegetation enriched with species encouraged for their economic value. Many of the "virgin" forests of the Amazon are old regrowth from previous cycles of slash-and-burn agriculture. And large swathes of those forests have been "enrichment-planted" with useful plants by indigenous groups and peasants, thereby transformed into *cultural* forests. None of the indigenous cultures along the Amazon cleared forest to raise livestock. Today, cattle and water buffalo ranching are the main driving forces behind deforestation on the Amazon floodplain. The sophisticated civilizations that once farmed the rich soils of the Amazon floodplain are testament to the potential of this environment for agricultural development without harm to the natural resource base.

Chapter 3 introduces the wild plants of the Amazon floodplain, examining their contribution to the diet and income of floodplain residents. The chapter is organized by use category. It begins with a discussion of edible plants, first those used for human consumption and then those used as bait to catch fish and turtles. A third large use category is filled by the wide variety of plants used for construction, household supplies, and crafts. Plants are also collected for a fourth major purpose: energy production; as the forests shrink, fuelwood supplies are fast emerging as a resource issue. Fifth is medicinal plants. Tropical forests are well known for an abundance of medicinal plants, and the Amazon floodplain is no exception. (Only a few of the medicinal uses of floodplain plants are described here, however, to illustrate the range of therapeutic concoctions rather than provide an exhaustive survey.) Finally, the chapter briefly reviews the history of the timber industry along the Amazon and discusses future prospects for it.

To devise appropriate development and conservation plans for the Amazon floodplain, it is essential to understand the driving forces that propel cattle and water buffalo ranching along the Amazon. Chapter 4 examines the history of ranching in the region, making the point that the impact of cattle raising on the Amazon floodplain has been minimal in most areas for centuries. Only recently have extensive tracts of floodplain forest been cleared to encourage pasture formation. The chapter explores the reasons for this dramatic turn-around in land-use, particularly in light of the region's rapidly growing urban population. Water buffalo did not make their appearance in the Amazon until the close of the nineteenth century; historically they have been largely confined to the seasonally flooded savannas of Marajó Island in the Amazon estuary. But water buffalo have now "broken out" of the estuarine area, making their way up the Amazon to affluents in Peru. Chapter 4 continues with a discussion of adoption patterns of water buffalo raising and an exploration of their environmental implications. It is not my purpose in this chapter to take a "position" on cattle or water buffalo, nor to suggest that they do not "belong" on the Amazon floodplain or elsewhere. Both small and large producers alike are responding to powerful economic and cultural motives when they take on cattle and water buffalo, or increase their herd size. Cattle and water buffalo help sustain the operations of small and large producers. Rather, I sound my underlying theme of a balanced approach to land use here, and emphasize the need to find viable economic alternatives to cattle and water buffalo production.

Chapter 4 concludes with a discussion of smaller livestock, which provide a significant source of income and subsistence for floodplain dwellers while producing minimal environmental damage. Pigs are kept year-round in elevated pens on the Amazon floodplain and are fed an assortment of fruits from flooded forests and home gardens. Muscovy ducks also stay on the floodplain throughout the year, while goats and sheep are transferred to upland pasture at high water. Many of these "micro-livestock" exhibit remarkable variation in breeds and the markets for their products are growing. Some wild animals that are adapted to the Amazon floodplain—and therefore do not require forest clearing to create suitable food resources—could eventually provide viable alternatives to cattle ranching. Several farmers have already taken the initiative in domesticating the capybara, the world's largest rodent, whose succulent meat is widely appreciated in rural and urban areas of the Amazon. A prolific breeder, the pig-sized capybara is at home on the Amazon floodplain, where it fattens on native grasses and other aquatic vegetation. Because Brazilian law forbids the raising of wild animals in captivity, the pioneering capybara farmers must conduct their operations in secret. It is ironic that society allows the destruction of biodiversity-rich forest to raise large Old World livestock, but does not permit the domestication of indigenous animals for meat production. Efforts are also being made to raise river turtles for urban meat markets, and the potential for domesticating other wild animals found on the Amazon floodplain, such the Amazon bamboo rat, are discussed.

Chapter 5, which discusses annual cropping, begins with a narrative of the social and economic drama triggered by the collapse of jute markets as a result of the introduction of cheaper synthetic sacking material. The closing of jute factories along the Amazon in the 1980s put thousands of people out of jobs, exacerbating urban poverty and forcing jute farmers to seek alternative crops or other livelihoods. Some former jute growers abandoned farming altogether, seeking out new lives in the burgeoning cities and towns along the Amazon and quickening the already powerful current of people moving from the countryside to urban centers.

Cattle ranching and increased attention to fishing for market were the most common options for those who remained on the floodplain. Cattle ranching, which has accelerated deforestation, has undercut an important food resource for many species of fish: fruits and nuts in the floodplain forest. As a result of habitat destruction and increased catches, some fisheries are on the verge of collapse. The post-jute scramble for alternatives thus has major implications for biodiversity, in both forest and field.

Cultivation of annual crops long grown on the Amazon floodplain, such as certain vegetables, is being intensified. The rapid expansion of farms engaged in the production of tomatoes, bell peppers, cabbage, and lettuce raises concerns about contamination of water—and eventually fish—with pesticides and fungicides. For the most part, commercial growers have adopted varieties of vegetables developed for temperate climates; in the hot and humid conditions of the Amazon, even more chemicals are needed to protect the delicate leaves and fruits from disease and insect attack. Other vegetable crops, on the other hand, are still dominated by traditional varieties and require few if any pesticide applications. The remarkable diversity of squash varieties still grown on the Amazon floodplain is testament to the fact that it is possible to maintain "heirloom" varieties in situ when markets can be found for their products.

Some see the vocation of the Amazon floodplain as that of a major granary, not just for Brazil, but for the world. But endless fields of mechanized rice production would be just as disastrous, and potentially unsustainable, as the proliferation of pastures currently under way. Chapter 5 also reviews the history of maize and rice cultivation, and current efforts to intensify production. Results with mechanized rice and maize cultivation on the Amazon floodplain have been mixed. In some cases, yields are no better than with traditional varieties using little or no purchased inputs. Opportunities for modest expansion of cereal production are analyzed within the framework of a balanced mix of land-use systems.

When the Amazon floodplain is mentioned in terms of becoming a major food producing area, the image of rice, and to a lesser extent, maize typically comes to mind. But manioc—the "poor man's" crop—is the main source of calories for inhabitants of the Amazon floodplain, as in many other parts of the region. Farmers have selected dozens of cultivars of both sweet and bitter manioc, one of the world's most productive root crops, which can be harvested within 6 months. Some of the manioc varieties are only grown on the floodplain, whereas others are cultivated on both uplands and seasonally flooded soils. Still other varieties are confined to *terra firme* because of their protracted maturation period. Farmers who live close to upland bluffs cultivate a "basket" of upland and floodplain varieties. Those farmers who live year-round on the wide floodplain, and do not have easy access to upland varieties of manioc, have honed their farming operations by focusing on varieties adapted to seasonally-available habitats, mainly the banks of water courses.

Manioc, then, exemplifies farmers' fine-tuning of agronomic practices and their keen abilities to mold agrobiodiversity to the exigencies of a highly variable environment. It has been remarked that whoever grows "poor man's" food will also become impoverished. Yet the market for manioc flour is assured and growing. Furthermore, manioc is the quintessential small farmer's crop and therefore an important asset in helping improve rural incomes.

Chapter 6 discusses tree farming on the Amazon floodplain. Perennials are about the last crop candidates that one would normally expect to deploy in flood-prone areas. Annual floods, it is argued, render tree crop farming virtually impossible. All farming environments have advantages and risks, however, and the Amazon floodplain is no different. Contrary to popular perception, numerous economically valuable tree and shrub species are well adapted to the varied flooding regimes of the Amazon. Higher banks of the Amazon may be covered by water only once every ten years or so, while lower-lying areas are typically under water for several months each year. Tidal influences create daily fluctuations of water levels along the lower Amazon. Rather than posing an insurmountable barrier to cultivation of perennials, the varied topography and tidal patterns of the Amazon provide a rich assortment of opportunities for farming. Farmers have learned to exploit these microenvironments with consummate skill. Orchard farming with diverse fruits, nuts, and other useful plants has been a significant economic activity along the Amazon in the past and still is in certain parts, but agroforestry has lost ground to jute and more recently to cattle and water buffalo production.

Agroforestry is one of the most biologically diverse agricultural production systems on earth and its environmental benefits have been well documented. But its potential for the Amazon floodplain remains largely untapped. Chapter 6 also examines the current contribution of agroforestry to the lives and economy of residents along the Amazon by focusing on the role of home gardens. Dooryard gardens are an overlooked resource for agricultural development: they are important recruiting grounds for new crops from the forest and contain many varieties that are preadapted to periodic flooding. Home gardens contain a stock of "protodomesticates" that are currently used as fish bait but could provide a lucrative source of feed for some fish-farming operations. Floodplain orchards could help restore some balance to land-use patterns along the Amazon, provided that sufficient markets and supporting infrastructure are in place.

Chapter 7 explores the constraints to more "environment-friendly" agriculture along the Amazon and suggests ways to enhance biodiversity in crop and livestock production by tapping into emerging market opportunities. The chapter discusses biases in the fiscal and regulatory environment and explores issues such as property rights, credit availability, and the degree of organization of growers. All stakeholders, from independent small farmers to communities and large-scale ranchers, need to be involved in adjusting the policy environment and devising plans for agricultural intensification. The intimate natural history knowledge of locals related to crops as well as wild and semi-domesticated plants and animals is a rich foundation upon which to build.

In temperate zones, flooding such as that in the Amazon floodplain is generally perceived as a threat to the livelihoods of residents. Hundreds of millions of dollars are spent "taming" rivers that once had broad floodplains, such as the Mississippi, and millions more are spent mopping up after the rivers eventually break through the dikes. The Amazon, which has never been forcibly channeled through an artificial course, should never be. Successful development of the Amazon floodplain hinges on taking advantage of the annual gift of fresh nutrients and silt, not on trying to prevent "harmful" floods. The harnessing of the Amazon for its hydropower would destroy the intricate cultural and ecological interactions along its generous floodplain.

2

The Ebb and Flow of Cultures

How many people inhabited Amazonia at the time of first contact with Europeans? More importantly to this discussion, what was their impact on the environment, particularly on biodiversity? Answers to these two questions can tell us a lot about the region's potential for sustaining population growth and development. We can also derive lessons from the legacy of natural resource exploitation during the colonial period, as well as from the boom-and-bust cycles over the last century or so.

The Amazon floodplain has often been considered sparsely settled, or inhabited only briefly by advanced cultures that arrived in the region and then soon declined. The region has thus been portrayed as largely empty, little touched—until recently—by human activities. Yet nearly a century ago, O. F. Cook, an ecologist with the U.S. Department of Agriculture, pointed out that few forests in the New World can be considered virgin.[1] Dense populations in many tropical forest regions of the Americas have resulted in substantial transformation of woodlands.

After introduced diseases and slaving raids into Amazonia decimated the indigenous populations, the forest returned in many cases, but with subtly different plant communities. Indeed, the Amazon forest, including many parts of the Amazon floodplain, has waxed and waned for thousands of years with the coming and going of cultures. Far from being pristine, the forests of Amazonia have been altered for millennia by slash-and-burn farming and deliberate enrichment by indigenous groups. And the tradition continues today with surviving aboriginal and peasant populations.

Amazonia has not been an insignificant backwater in the cultural history of South America. Rather, it has been an important region for cultural adaptation, innovation, and crop domestication. Furthermore, human populations were much denser in precontact times than estimates in the literature typically grant. By the time Europeans arrived, in the early sixteenth century, the Amazon floodplain and its adjacent uplands were far more densely settled than they are even at present. Yet dense settlement in the wetlands of the Amazon and bordering *terra firme* did not disrupt the natural resource base, nor trigger deforestation on the scale evident today. This chapter traces the ebb and flow of Amazon cultures from the first hunters and gatherers through the first farmers, then examines the environmental implications of dense indigenous populations. Land-use dynamics changed profoundly with the devastating population crash that attended the region's first contact with Europeans; the chapter traces those changes from the colonial period to the present day.

The First Arrivals: Hunters and Gatherers

The length of human habitation in the Americas, let alone the Amazon floodplain, is still a matter for scholarly dispute. While an outside date of 10,000 years has been suggested in the past, more recent evidence suggests that hunters and gatherers penetrated diverse environments in the Americas at least 30,000 years ago.[2] Stone tools at the Pedra Furada (perforated rock) rock shelter in Piauí in northeastern Brazil were apparently employed as far back as 48,000 years ago. To arrive in Piauí (figure 2.1), early foraging groups would have mostly likely passed through Amazonia, or skirted its eastern fringe along the coast.

While some scholars contest these early dates, other lines of investigation support a long history of human occupation in the Americas. Analysis of the DNA of aboriginal groups in various parts of the Americas points to a possible first wave of migrants reaching the New World from Asia some 42,000 years ago. Linguistic evidence suggests that people first entered the New World some 50,000 to 60,000 years ago if only one migrant group is the founding stock for indigenous languages of the Americas, and some 30,000 to 40,000 years ago if migrants arrived in waves.[3]

The Amazon basin was inhabited long before the advent of open-field agriculture. Waves of hunters and gatherers probably penetrated Amazonia at various times from different directions. Convention holds that people settled South America by traveling down the Andean mountain chain from Central America, avoiding the rainforests to the east. Only later are people thought to have ventured into the "inhospitable" jungles of the Amazon and the upper Orinoco. Civilization and antiquity of settlement in South America has long been strongly identified with South America's backbone mountain chain.

But there is no reason to believe that people did not fan out into Amazonia soon after they arrived in South America. The rich forests of Amazonia, stocked with game and fruits, and the generous floodplains full of fish, turtles, and other aquatic animals, offer a more ample banquet than does the land around the Andes or the Pacific coast of South America, much of which is a harsh desert. The additional notion that people only tentatively probed Amazonia by sticking close to open habitats such as the savannas of Roraima and Amapá is also without merit. While the climate may have been drier than it is today in Amazonia when some

Figure 2.1. States within Brazil.

groups entered the region, they would not necessarily have clung to fingers of savanna. The Isthmus of Panama, through which migrans to South America passed, was largely forested from 14,000 to 10,000 years ago. While grasslands and scrub woodland in Amazonia are not devoid of game, they are not stocked with vast herds of ungulates to attract settlers. Therefore, the idea that savannas are more attractive to hunters and gatherers is without merit.

Nor did early waves of immigrants necessarily follow river courses. It seems just as likely that they would have trekked through interfluvial forests. Some may have lingered in upland forests, rather than have settled along rivers.

Many groups of hunters and gatherers undoubtedly built base camps close to, or even on, the floodplains of certain rivers, especially the Amazon. Some groups may have built their camps on the higher parts of the floodplain, or even in the trees of seasonally flooded forest, because they specialized in foraging in such environments.[4] Ultimately, however, the question of whether hunters and gatherers lived on the Amazon floodplain is unlikely to be resolved. The repeated rising and falling of sea level during the Pleistocene has eroded, then buried, any cultural remains. Sea level dropped by at least one hundred meters during the height of the last Pleistocene glaciation, triggering pronounced scouring of river channels. When sea level started rising again, as the ice sheets melted around 15,000 years ago, the Amazon became partially backed up, dropping much of its sediment load and forming a broad floodplain in the process. Fresh deposits of alluvium on such a large scale essentially redesigned the floodplain. The large bay-mouths of certain tributaries of the Amazon, such as the Tapajós and Anapú, now resemble the rias of the formerly glaciated valleys of Norway, an effect achieved by successive deepening of their channels followed by the backing up of their waters as sea level rose.

The interface between upland and wetland environments in the Amazon floodplain offers the richest array of food resources to support foragers. And campsites on upland bluffs would not have been affected by the fluctuating sea levels. The towering sandstone hills overlooking the floodplain just west of Monte Alegre are one such favorable environment (figure 2.2). Serra do Pilão (mortar-shaped hill), a hill bearing pictographs, would have been a superb site for hunters and gatherers because it overlooks fish-rich Lago Grande de Monte Alegre, a lake so large one could easily imagine an open sea. Cave and rock-wall paintings still adorn some of the exposed boulders, depicting geometric designs, frogs, mammals, figures of humans or spirits, and handprints (figure 2.3). By deploying various dating techniques on cultural remains at one cave, Pedra Pintada (painted rock), near Serra do Pilão, Anna Roosevelt and her group have established that the site was first occupied about 11,200 years ago. One date from cultural remains at the Pedro Pintada site raises the possibility that people first arrived there 16,000 years ago.[5] The majority of the intriguing pictographs at Pedra Pintada were likely painted between 11,730 and 9,880 years ago. A varnish-like coating of silica has protected the iron oxide pigment from the fierce tropical sun and torrential rains.

The significance of the rock paintings is unclear, but some of them probably represent visions of shamans while under the influence of hallucinogenic compounds. Early stages of trances induced by hallucinogenic plants allegedly conjure images such as zigzags, grid patterns, dots, and concentric circles, and such images are depicted at Pedra Pintada. The groups responsible for the imaginative designs clearly had plenty of spare time to procure pigments and paint. They were obviously not on the brink of starvation. On the contrary, they were likely well-fed, with a rich spiritual life. Knowledge of plants used in curing and mediating with the spirit world predates agriculture by a wide margin. The earliest domesticated plants were used for treating the sick and dealing with the supernatural, rather than for producing food.

What resources did the artistic groups at Pedra Pintada draw upon to achieve and support a lifestyle that included plenty of time for leisure and spiritual contemplation? Fruits and nuts from several upland forest trees have been found at Pedra Pintada, including Brazil nut, jutaí, and pitomba. The fruits of tucumã palm (appendix A) have also been encountered at the site. This palm is never found in mature upland forest (figures 2.4, 2.5). Rather, it is typical of disturbed sites, suggesting that the inhabitants of the Pedra Pintada cave may have encouraged the productive palm by protecting spontaneous seedlings and partially clearing and burning the forest. Remains of tarumã fruits, a floodplain tree, have also been found.

Animal bones and shells reveal that the groups foraged for game in both upland and aquatic environments. By living along the interface between the uplands and floodplain, early settlers clearly had the best of both worlds. Animal protein sources included various species of fish, mammals, turtles, tortoises, and freshwater pearly mussels.

Figure 2.2. Sandstone hill overlooking the floodplain where hunters and gatherers lived at least 11,000 years ago. Early settlers exploited aquatic resources as well as upland game and forest fruits. Serra do Pilão near Monte Alegre, Pará, March 1993.

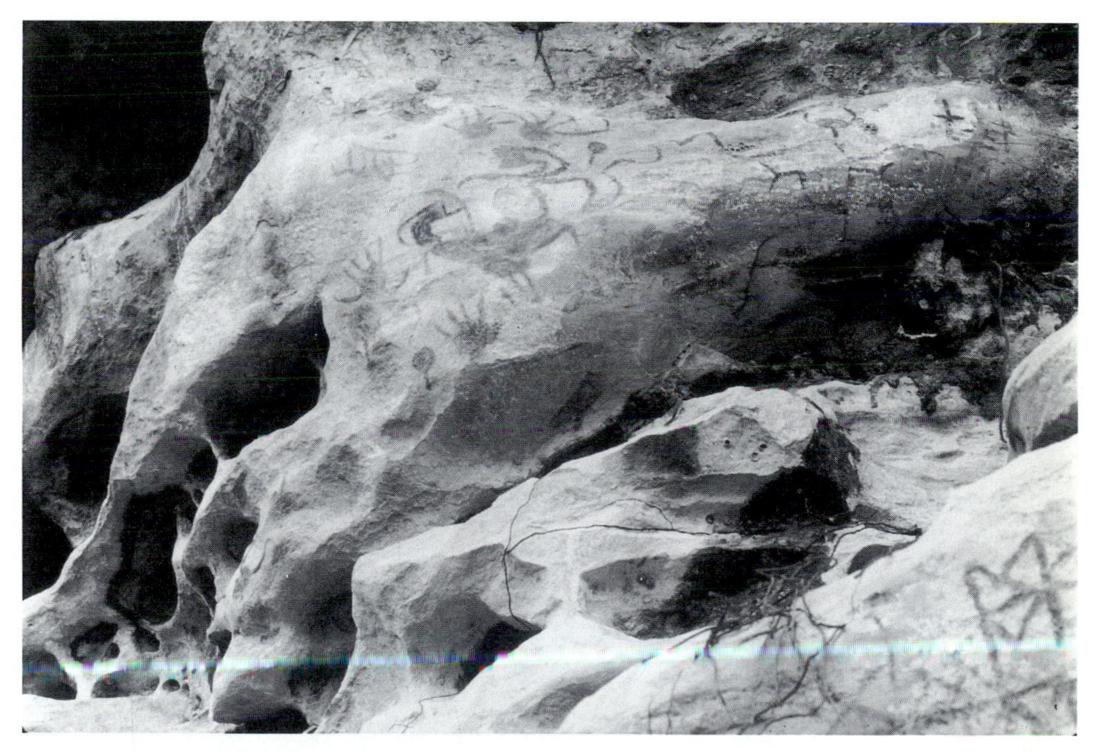

Figure 2.3. Rock paintings, applied with ochre and yellow paint some 11,000 years ago. The meaning of the ancient paintings is obscure, but some may represent images perceived under the influence of hallucinogenic compounds. Others may depict sacred objects, or may have served as devices to mark the passage of seasons. A large handprint is discernible in the center of the photograph. Pedra Pintada near Serra do Pilão in the Municipality of Monte Alegre, Pará, March 1993.

Figure 2.4. Tucumã palm (*Astrocaryum vulgare*) with fruit. Tucumã palms occur in disturbed sites and are an indicator species for former human occupation of an area. Lower Maracá River, Amapá, May 1996.

Figure 2.5. Vitamin A-rich fruits of the prickly tucumã palm (*Astrocaryum vulgare*), gathered from a home garden on the margin of the floodplain. These palms are occasionally planted in home gardens on higher parts of the floodplain. In non-flooded upland areas, the palm is usually spared around homes because of the fruits and the tough fiber that can be obtained from the fronds. Santana do Ituqui, Pará, March 1993.

Shell mounds have been found at various sites on or adjacent to the Amazon floodplain and the lower reaches of some of its tributaries, further testament to the rich foraging grounds afforded by living along the border between uplands and wetlands. Taperinha is one such site, across the Amazon from Pedra Pintada (figure 2.6). Anna Roosevelt and her group have found the oldest pottery thus far from the New World at Taperinha, dating it between 8,025 and 7,170 years ago.[6] Taperinha borders the Amazon floodplain along a channel that drains a backswamp lake that also receives sediment-rich water from the Amazon. The shell mound at Taperinha covers several hectares and is six meters deep. Hunters and gatherers, not farmers, were primarily responsible for these shell mounds. Surprisingly, the several species of freshwater mussel that provided one of the main sources of animal protein for the ancient pottery makers of Taperinha are no longer harvested in the area. When and why shellfish became a less significant food item is not known, but by the late nineteenth century few people in the area were bothering to collect and eat them.

The shell mound at Taperinha was partially excavated in 1969 by the Jari pulp operation as a source of calcium and phosphorus fertilizer for recently established plantations of Gmelina, Caribbean pine, and eucalyptus in northeastern Amazonia. Shell mounds along the Amazon and some of its tributaries have been mined for centuries to obtain lime for building purposes. Shell mounds reportedly occur along the middle Amazon, such as at Porto do Itã around the margins of Lago Grande da Franca, between Santarém and Juriti. Shell mounds near the mouth of

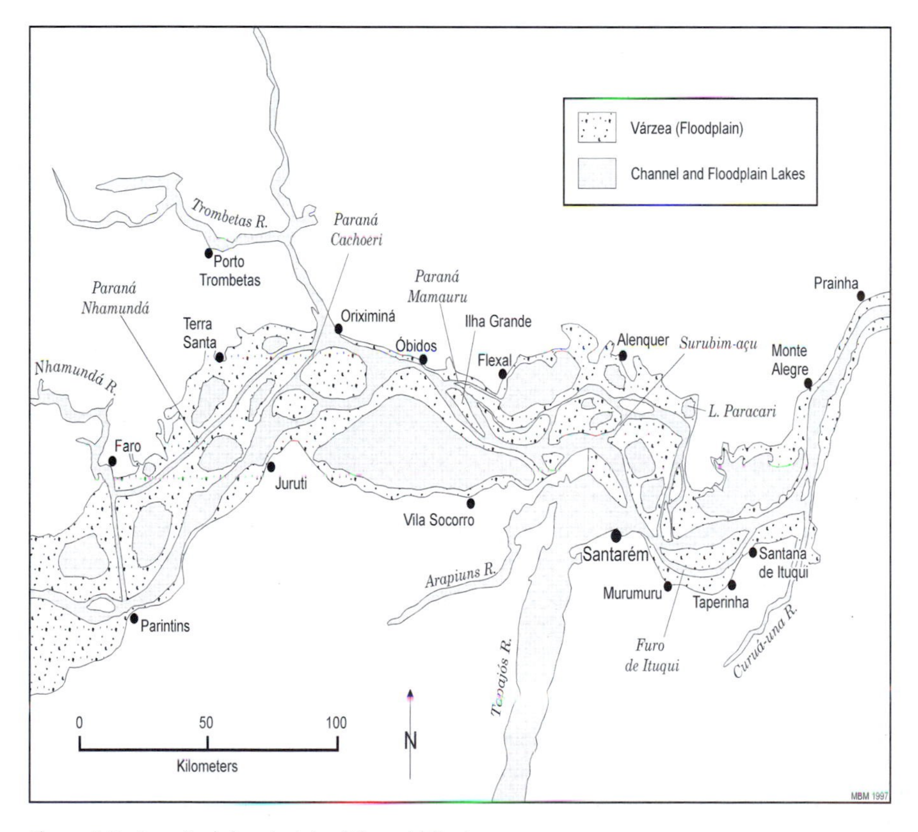

Figure 2.6. Faro-Prainha stretch of the middle Amazon.

the Trombetas, a northern tributary of the Amazon, were mined to help build the hilltop fort at Óbidos, which was established in 1697 and rebuilt in 1854, and overlooks one of the narrower stretches of the middle Amazon. An industry once flourished at Cametá along the lower Tocantins making *cal* from crushed shells taken from mounds on the banks of the clear-water river. The calcium-rich powder was mixed with water to paint buildings and stirred into mud to make adobe bricks. Relatively intact shell mounds can still be found at the mouth of the Anapú, a bit off the beaten track, which has saved the shell mounds from devastation.

Communities engaged in hunting, gathering, and fishing established themselves long ago in other parts of the Amazon Basin. Two stone arrow heads from the middle Tapajós are presumed to have been fashioned by hunters and gatherers some 10,000 to 8,000 years ago. Preceramic refuse containing percussion-flakes and remains of plants and small animals, found in a rock shelter in the Carajás range in southeastern Amazonia, are about 8,000 years old. It is clear that foragers dispersed far and wide throughout the 7-million-square-kilometer Amazon Basin; they did not shun it because it lacked grassy plains stocked with herds of deer, bison, or mammoth.

The Emergence of Farming

Long before the first Europeans came to Amazonia in the early 1500s, prosperous chiefdoms had been established along extensive stretches of the Amazon. Sizable and well-organized societies had learned to farm both the Amazon floodplain and its adjacent uplands. Following in the footsteps of their hunter-gatherer forebears, farmers managed natural resources skillfully on both *terra firme* and wetlands. Trade networks were well developed, and agriculture was most likely intensive in some cases, especially along the Amazon.

Some anthropologists, particularly Betty J. Meggers, have argued that the tropical rainforests of Amazonia were incapable of supporting dense human populations, even after the advent of farming.[7] Others who have dismissed environmental determinism as an "explanation" for the lack of indigenous states in Amazonia, have themselves argued that agriculture was never intensified because population pressure did not trigger much agricultural innovation.[8] I reject both notions.

Four main assumptions underlie my examination of agricultural history and human-carrying capacity along the Amazon. First, the environment was not inimical to the development of higher cultures. Second, populations became dense along many parts of the Amazon floodplain and adjacent uplands. Third, people intensified agriculture, especially as population pressure built up. Fourth, reports of early explorers about the large size of indigenous groups cannot be dismissed outright as untrustworthy, as did Meggers.[9] In Mexico and Central America, for example, Sherburne Cook and Woodrow Borah have shown that accounts by Spanish explorers of large indigenous settlements and large-scale transformation of the landscape at time of contact are supported by archaeological and historical evidence.[10] A similar picture is emerging in Amazonia.

By the time Europeans arrived in the New World, much of the uplands bordering the Amazon River were densely settled. At the mouth of the Tapajós, for example, over two hundred canoes containing twenty to thirty warriors each met Fran-

Figure 2.7. Potsherds found in a backyard in Santana do Ituqui, an upland village at the edge of the floodplain. Santana do Ituqui is a black-earth site with numerous pottery fragments, many with zoomorphic figures. The young girl's physical features suggest links to indigenous ancestors who first occupied the area thousands of years ago. Near Santarém, Pará, March 1993.

cisco Orellana's expedition in 1542. Orellana was a Spanish explorer who started in Quito and entered the Amazon by way of the Napo River. On shore, more warriors and villagers shouted defiance. With some 5,000 warriors in the water, and at least an equal number on shore, the Tapajó population must have exceeded 40,000.

Some discount the reports of early explorers with respect to the size of indigenous populations. Expeditionary forces had a vested interest in inflating the size and wealth of the indigenous populations in order to whet the appetites of sponsors in their European homelands. But in the case of the Tapajó village, now the site of Santarém, we have more than early eyewitness accounts. The main village of the Tapajó extended for several kilometers, as evidenced by anthropogenic black earth covering substantial parts of the older districts of Santarém. (Anthropogenic black earth is formed by the accumulation of kitchen middens as well as ash and charcoal from fires kept burning continuously for cooking and warmth at night.) The Tapajó apparently lived in several satellite villages stretching along the upland bluffs. Deep black earth, rich in potsherds and the occasional stone axe, occurs virtually uninterrupted for some thirty kilometers downstream from Santarém to Santana do Ituqui (figure 2.7).

While evidence suggests that at least the Santarém area was densely settled in 1542 when Orellana's expedition hurried by, what about the rest of the Amazon River? Introduced diseases often take a quick toll on the unimmunized, so the most realistic picture is likely to be revealed in the earliest accounts. The first Europeans to descend the Amazon were part of Orellana's expedition.[11]

A priest, Friar Gaspar de Carvajal, accompanied Orellana on his journey down the Amazon and wrote an account shortly after the expedition.[12] According to Carvajal, several stretches of the uplands bordering the Amazon, not just the Santarém area, were densely settled in 1542. And what of the "empty stretches"? Orellana and his crew descended the Amazon during the flood season. Parts of the floodplain that were likely intensively farmed at low water were thus under water and appeared unoccupied. Because Carvajal reported "empty stretches" along the Amazon floodplain, some historians and anthropologists, such as Warren DeBoer, have seized on this observation to argue that the Amazon River contained substantial buffer zones between powerful chiefdoms.[13]

The notion of buffer zones sounds logical, but is misleading. Although a no-man's-land likely existed between the Omagua, a chiefdom of some seven hundred kilometers along the Upper Amazon, and the Machifaro, who were based around the present-day town of Tefé, other supposedly empty stretches of the river were likely inhabited. Orellana stuck to the main course of the Amazon and probably remained midstream where the current is swift. Along some stretches of the river, such as between Nhamundá and Oriximiná, a succession of lakes and interconnecting channels extends back some thirty kilometers before reaching the uplands. Many upland bluffs could have been densely settled without Carvajal realizing it because they would have been out of sight. Orellana wisely avoided taking narrower side channels where the expeditionary force would have been vulnerable to surprise attack, or could have become lost. At high water, floodplain lakes often merge into one another, and a boat without a knowledgeable pilot can easily run aground when it encounters a submerged bank.

It is known that Orellana's expedition often had a hard time locating food supplies. The Spaniards' difficulties might be taken to imply that the population around the Amazon was not particularly dense and that food was in short supply. But because the Orellana expedition quickly earned a reputation for treachery, as well as for pillaging and sacking villages, indigenous groups understandably retired with their food, or attacked the expedition as it descended. Well into the early colonial period, ships avoided the north bank of the Amazon between the Negro and opposite the confluence with the Madeira because of the strong Indian presence which had learned not to trust Europeans.

On the few occasions that Orellana's expedition came ashore, Carvajal noticed that substantial trails often led inland. Such well-worn paths probably led to cultivated plots and other settlements. The Tapajó, for example, walked along paths leading into the interior that were about a meter and a half wide and had apparently been worn down and eroded some thirty centimeters. Not only were many of the upland bluffs occupied by villages stretching for several kilometers in some cases, but some inland areas were also densely settled.

Evidence in the form of black earth and raised mounds supports the idea of dense settlements not only on the floodplain but also on upland bluffs and inland areas. The anthropogenic nature of black soils in the Amazon was recognized over a century ago. As C. Barrington Brown and William Lidstone commented in 1878:

> Villages must have stood upon these spots for ages, to have accumulated such a depth of soil about them . . . At the present day these localities are highly prized as agricultural grounds, owing to their fertility; and they bear the name

of "Terras pretas" (black earths). We have observed them occurring in many places almost too numerous to mention.[14]

Anthrosols, as these soils are sometimes termed, are sufficiently common in the Brazilian Amazon to be recognized as a distinct soil type: *terra preta do índio*. Anthrosols are formed on a variety of soil types and are characterized by phosphorus levels much higher than the surrounding soils, undoubtedly from inclusion of fish and game bones in kitchen middens and human wastes. Phosphorus levels of Amazonian anthrosols vary, but are always higher than the surrounding soils, the vast majority of which contain phosphorus in levels lower than 10 parts per million (ppm). The phosphorus level in one black-earth site overlooking the Amazon floodplain, at Fazenda São Jorge near Ipaupixuna along the road to Murumuru road in the Municipality of Santarém, for example, was 184 ppm. The anthrosol at the entrance to Pedra Pintada cave had a phosphorus level of 339 ppm.

Black-earth sites associated with potsherds are common on uplands adjacent to the Amazon floodplain or close to it (table 2.1). Anthrosols are especially no-

Table 2.1 Some anthropogenic black-earth sites adjacent or close to the Amazon floodplain between Manacapuru and Monte Alegre, Brazil

Location	Depth (m)	Area (ha)	P (ppm)
Juriti, Pará	1.2	350.0	147.0
Comunidade Terra Preta, Igarapé Cacoal, km 55 of Juriti-Tabatinga highway, Pará*	1.0	200.0	134.0
Manacapuru, Amazonas	1.0	80.0	33.0
Terra Nova, Paraná do Silves, 1 km W. of Itapiranga, Amazonas	0.9	45.0	155.0
Itacoatiara, Amazonas	1.0	30.0	120.0
Fazenda São Jorge, Ramal Murumuru, near Santarém, Pará	1.0	20.0	184.0
Itapiranga, Paraná do Silves, Amazonas	1.0	8.0	139.0
Km 4 of Cacau Pirêra–Manacapuru highway, Amazonas*	0.8	4.0	8.0
Fazenda Paraíso near Farol, Pará	0.8	2.0	97.0
Campina, estrada de Terra Preta, near Cacau Pirêra, Amazonas*	0.2	1.0	0.2
Ceramica Irco, km 15 of Aleixo road, Manaus, Amazonas	0.5	0.8	38.0
Carariacá, Pará	0.7	0.5	312.0
Lago de Madruba near Itapiranga, Amazonas	0.5	0.5	15.0
Estrada de Terra Preta near Cacau Pirêra, Amazonas*	0.2	0.5	0.3
Lago de Terra Preta, 30 km S.E. of Itacoatiara, Amazonas	0.8	0.4	33.0
Entrance to Lago de Batista, 25 km S.E. of Itacoatiara, Amazonas	0.4	0.4	35.0
Pedra Pintada, Serra do Pilão, near Monte Alegre, Pará*	0.3	0.2	339.0

*Denotes inland site a few kilometers from the floodplain of the Amazon River, but occupants likely exploited resources on the floodplain, at least seasonally.
Source: From Smith, N. J. H., "Anthrosols and Human Carrying Capacity in Amazonia," *Annals of the Association of American Geographers* 70 (1980), pp. 553–566, field sampling 1992–1994.
Notes: Soil samples analyzed by CPATU-EMBRAPA and CEPLAC, courtesy of Italo Falesi. The anthrosol at Campina is located on a podzol, which may account for its low phosphorus level.

ticeable along the Manaus-Manacapuru section of the Amazon, in the vicinity of Itacoatiara, and from Nhamundá to Monte Alegre. The black-earth site at Santarém covers five square kilometers, and half a century ago, Curt Nimuendajú recorded sixty-five other anthrosols in the vicinity of the confluence of the Amazon and Tapajós.[15]

One of the larger anthrosols along the Parintins-Santarém stretch of the Amazon is at Juriti. Today, Juriti is a small riverside town of about 15,000 inhabitants, but the site was more densely occupied in the past. Juriti is located on a 350-hectare expanse of anthropogenic black soil that extends beyond the present-day limits of the urban area. The potsherd-laden black earth stretches for about 3.5 kilometers along the waterfront and about 1 kilometer inland. Dense groves of mucajá palm on the outskirts of town are also a sign that people have settled in the area for some time (figure 2.8). The black earth at Juriti is a favorite with present-day vegetable and tobacco growers.

A 200-hectare black-earth site full of potsherds at km 55 of the Juriti-Tabatinga road is just as impressive. This anthrosol is 1 meter deep, suggesting that the site was occupied by a large village for an extended period. Locals recognize the exceptional fertility of the site, and have named their community after it: Comunidade Terra Preta. One indication of the abundant soil nutrients in the anthrosol at Terra Preta is the abundance of papaya that have sprouted spontaneously from seeds dropped by birds (figure 2.9). Although Terra Preta is on an interfluve

Figure 2.8. A grove of mucajá palms on a black-earth site currently in pasture. Mucajá palms, found in disturbed sites, are an indicator species for former human occupation of an area. The oily fruits of the palm serve as a snack food, and are also collected to feed pigs. Outskirts of Juriti, Pará, June 1994.

Figure 2.9. Spontaneous papaya growing on a black-earth site of 200 hectares that has been partially cleared to cultivate manioc. The young girl in the center is holding potsherds that were picked up on the soil surface, indicating that the site was once occupied by an Indian village. Comunidade Terra Preta, km 55 of the Juriti-Tabatinga road, June 1994.

between the Amazon and Arapiuns, people surely harvested resources from the Amazon floodplain some fifteen kilometers away in ancient times. Today, inhabitants of the Arapiuns watershed often fish in the Amazon, because fisheries in the black-water Arapiuns are relatively poor. The former inhabitants of Terra Preta probably established their village there because the best sites along the upland bluff overlooking the Amazon were already occupied. Farmers recognize the superior fertility of the numerous black-earth sites in the Arapiuns watershed and cultivate manioc on them to make flour for subsistence and commerce.

Black-earth sites are less common on the Amazon floodplain itself, and William Denevan, a geographer, accordingly speculates that settlements were probably small and sparsely distributed.[16] The periodic floods are perceived as inimical to permanent and dense settlement on the floodplain. But as Daniel Lathrap recognized, one cannot ascertain population density in the past based merely on the surviving archaeological sites, because most ancient village sites on the floodplain have been destroyed by the river which constantly reworks its channels.[17]

Enough black-earth sites survive on the Amazon floodplain to suggest that at least some parts of the *várzea* were densely settled for a long time. The Brazilian geographer Hilgard O'Reilly Sternberg has recorded anthrosols on the higher and more stable parts of Careiro Island on the Amazon floodplain near Manaus.[18] At Urucurituba near Santarém, farmers uncover potsherds and stone axes when they cultivate black-earth patches on the floodplain near their homes.

At least one anthropogenic mound reportedly occurs on Carmo Island (also known as Ilha do Meio), about halfway between Santarém and Óbidos. Artificial mounds have been known for some time on Marajó Island, and the archaeologist Anna Roosevelt suggests that the population density there averaged at least five to ten persons per square kilometer in prehistoric times and may have been much higher.[19] Several farmers are aware that indigenous people constructed the mound on Carmo Island because it allegedly contains black earth and potsherds. Locals refer to it as a *teso* (hill). The mound is located well within the island, about two kilometers from Paraná do Meio. Because the mound is never flooded, farmers and ranchers have used it to corral cattle.

Uninhabited black-earth sites are thought to harbor spirits and are generally avoided after sundown. For example, farmers generally avoid the anthropogenic mound on Carmo Island at night because they say it is haunted. An old-time resident of Carmo Island, eighty-three years old at the time of our conversation in 1994 and now deceased, recounted a tale that happened to his cousin Arturo many years ago. Arturo tried to sleep one night at the mound, but a ghost kept pushing up on his hammock, trying to tip him out. The persistent poltergeist eventually forced Arturo to spend the night in a tree.

In addition to settlements on mounds and along higher banks of the Amazon floodplain, villages were likely built on stilts in the middle of some floodplain lakes. Mosquitoes can become almost unbearable at certain times on the Amazon floodplain, and residents today generally resort to burning arboreal termite nests, cattle dung, or leafy twigs to ward them off. Mosquito hordes drop off dramatically as one proceeds from the margin of a lake to its center. Another way to avoid mosquitoes would be to assemble villages on floating platforms in the middle of lakes. With the same idea, fishermen often resort to napping in their canoes in the middle of lakes while waiting for fish to become entangled in their gillnets. Although lakes are often vast, they are generally shallow, usually under ten meters even at high water. The only real danger of establishing villages in the middle of lakes is wave damage during storms. With lakes stretching for several kilometers in some cases, the long fetch would have allowed sizable waves to form during strong thunderstorms. The now-extinct Paumari of the silty Purus River once lived on floating villages in the middle of lakes at high water. It seems unlikely that the Paumari were the only cultural group to come up with that solution to troublesome mosquitoes. Villages on large floating logs or sturdy stilts were surely hidden from view in the interior of floodplain islands and in backwater lakes when Orellana descended the Amazon.

Estimates of the precontact population density of the Amazon Basin vary widely. Marajó Island alone may have had a million inhabitants when the Marajoara culture flourished from A.D. 400 to 1300. The population of the Amazon floodplain and adjacent uplands, including its main headwater rivers the Ucayali and Marañon in Peru, likely reached 10 million by the time first Europeans arrived. For Amazonia as a whole, then, the population was probably about 15 million in 1500.

Plenty of high-quality protein and productive crops were available to support such high population densities. Fish and turtles continued to be important items in the diet thousands of years ago as groups took up farming. Turtles were even kept in corrals like river cattle and large quantities of fish were sun-dried at low water to provision villages during the flood season when fish were harder to catch. Manatee, capybara, and aquatic birds would have also supplemented the diet, as they had since people first came to the shores of the Amazon river-sea.

Little is known about the prehistoric cropping patterns that supported large human populations, but sweet potato was probably an early basic staple. Sweet potato is ready for harvesting in a few months and may have been cultivated along the Amazon floodplain as long as 10,000 years ago. Convention holds that agriculture first "started" in the Fertile Crescent of the Middle East with the domestication of wheat and barley. But as Carl Sauer suggested, fisherfolk may have been the earliest farmers and they likely cultivated root crops.[20] Sauer pinpointed coastal Southeast Asia as the likely region where crop production first got under way, but the Amazon floodplain seems just as propitious a setting.

Cocoyam was likely another early root crop on the Amazon floodplain, because this aroid is native to the New World and is well-adapted to wet places. In addition, the New World yam might also have been cultivated on the higher parts of the floodplain as people began deriving more of their subsistence from farming. Neither cocoyam nor the New World yam appear to be important on the *várzea* today. Cocoyam remains have been found in raised fields in the Llanos de Mojo in the Bolivian Amazon. Some of the raised fields in the seasonally flooded savannas were built 2,000 years ago. The Llanos de Mojo are flooded by whitewater rivers, particularly the Mamoré and the Beni, and were once densely settled. Today, the New World yam and cocoyam are mostly confined to dooryard gardens in the Brazilian Amazon, particularly on the uplands.

Manioc was probably introduced a little later than other root crops to the Amazon floodplain. Manioc is typically harvested six to eighteen months after planting, and it would have taken time to select precocious varieties that can be harvested before the annual flood. Nevertheless, manioc is still an ancient crop on the floodplain, where it has most likely been cultivated for several thousand years. Initially, the crop would have been cultivated on the highest parts of the floodplain, which are not normally covered with water every year. Later, however, rapidly maturing manioc varieties would have been selected to take advantage of the much greater area of lower-lying floodplain soils. The Omagua used to bury manioc tubers in holes for the duration of the flood. After the waters receded, the tubers were dug up to prepare a thick pancake (known as *beijú* in Brazil) that could be stored for months without spoiling. The tubers did not rot in the well-tamped holes because of anaerobic conditions.

Given the plentiful supplies of fish, various species of turtle, manatee, capybara, and several species of ducks, particularly the black-bellied tree duck and the Brazilian duck, heavy reliance on starchy root crops would not have led to protein malnutrition. As late as the mid-nineteenth century, black-bellied tree ducks were so plentiful on parts of the Amazon floodplain, such as Marajó Island, that the horizon darkened when they took flight. Today, heavy hunting pressure has so thinned duck populations that they are encountered only in small groups.

Vegetable sources of protein would have also supplemented the diet. It is not known when beans were first cultivated on the Amazon floodplain, but kidney beans were domesticated in two parts of its extensive range, Mexico and Bolivia. It is conceivable that varieties developed out of the southern gene pool may have been grown on the Amazon floodplain at low water several thousand years ago.

Cereals are a good source of vegetable protein, but they appear to have been less important than root crops, at least initially. Extensive floating pastures of wild rice were once harvested by shaking the ripe panicles into canoes or baskets (figure 2.10). On Marajó, Anna Roosevelt suggests, some of the wetland grasses may have been domesticated, such as *Leersia hexandra*.[21]

Maize arrived in South America relatively early, but apparently only became a widespread and significant source of food late in the agricultural history of the continent. A highly productive and nutritious cereal, maize soon spread from its area of origin in Mexico and Guatemala, apparently reaching the Napo River in western Amazonia some 6,000 years ago, and the middle Caquetá 4,700 years ago. Maize moved down the Amazon where it may have been used mostly for making fermented drinks rather than for eating. In 1542, Orellana's expedition encountered a village along the Amazon between the Negro and Madeira with massive tree-trunks carved in the form of two jaguars supporting a tower with a central font containing maize beer. But dense indigenous populations along the Amazon did not hinge on the cultivation of maize. Root crops and a broad-spectrum diet based on extractive products from forests, fish, and game would have permitted large concentrations of people on the Amazon floodplain and adjacent uplands. Maize may well have been a minor crop for the precontact populations of the Amazon River, as it is today.

Some indigenous groups may have specialized as floodplain farmers, but most probably farmed both the floodplain and adjacent uplands, as numerous communities on and around the floodplain still do today. It also seems likely that trade networks were soon established between people living along the Amazon floodplain and more inland locations. Foodstuffs and other goods were probably bartered between disparate groups. Some villages located dozens of kilometers from the floodplain may thus have benefited from the rich natural resources of the Amazon River.

Environmental Implications of Dense Indigenous Populations

By the time the Europeans arrived in the early 1500s, the population density of the Amazon floodplains and the rural uplands was much higher than it is today. Only recently has the population of the region begun to approach its former numbers. The main difference is that today's population is mostly urban; almost two-thirds of the region's population lives in towns and cities. Although some indigenous civilizations along the Amazon had sizable villages, none rivaled the dimensions of the region's current megacities of Belém and Manaus.

In precontact times, the countryside was densely settled. Still, in spite of the large concentrations of indigenous groups on the Amazon floodplain and adjacent floodplains, the natural resource base was not degraded. Turtles, for example, were abundant until reckless exploitation took its toll in the colonial period. Several reasons account for the fact that the natural resources were not degraded, at least on any significant scale. Precontact indigenous people did not raise cattle or water buffalo. In the absences of natural savannas, cattle ranching requires extensive forest clearing. Furthermore, cultural checks were in place with many groups to prevent overzealous harvesting of natural resources, as spirit protectors of economic plants, game, and fish punished the greedy. Also, indigenous agricultural systems were undoubtedly diverse and intensively managed. Productivity was likely high, thereby reducing pressure on surrounding forest. Large

Figure 2.10. A meadow containing mostly wild rice on the floodplain. Although floating meadows are indigenous to the Amazon, they have expanded dramatically over the last few decades as a result of deforestation. Near the mouth of the Nhamundá River, June 1994.

populations along the Amazon floodplain could thus have been supported with far less forest-clearing than is common today.

Although appreciable forest remained standing at any one time in precontact times, little of it was virgin. At various times virtually all of the forest on the higher parts of the floodplain was likely cleared, farmed, and then allowed to revert back to forest, perhaps as a managed fallow for a decade or more. The floristic composition of floodplain forests has thus long been influenced by previous land-use practices and by artificial enrichment. Açaí and bacabinha palms, for example, have been planted along the banks of the lower Amazon. Bacabinha continues to be planted along the banks of the middle Amazon, particularly in home gardens, such as such as at Itapiranga, a village downstream from Itacoatiara in the state of Amazonas. Home sites with bacabinha and açaí may later be engulfed by forest reasserting itself after the homesite is abandoned. Such economic species thus blend in with spontaneous regeneration to create a cultural forest (figure 2.11).

Once introduced or artificially enriched on the Amazon floodplain, economic plants often reproduce on their own. Other fruit trees most likely introduced along the Amazon floodplain, or at least artificially enriched, include cacao and yellow mombim. Cacao is native to western Amazonia, where it occurs in upland forests as well as along the floodplain of white-water rivers such as the Purus. Large concentrations of cacao were reported along certain portions of the Amazon between Manacapuru and Coarí by Francisco Xavier Sampaio in the eighteenth century; the itinerant judge, an astute observer of the landscape and culture of the region, speculated that the cacao was "planted by nature."[22] But it is not clear

Figure 2.11. A cultural forest rich in açaí palm surrounding a couple of farmhouses along the lower Amazon. Cultural forests in this area have been established by planting perennials after short-cycle crops. Thus, what used to be the fallow vegetation between slash-and-burn cropping cycles with annual crops is now occupied by productive agroforestry systems. Furo de Breves, Pará, November 1995.

whether they were truly natural. Although cacao may be native to the floodplain of the Upper Amazon, its occurrence in such density is likely the result of human agency. In describing Aramaça Island along the Upper Amazon near the mouth of the Javarí in 1774, Sampaio notes that cacao was so abundant that the entire island was practically a plantation (*"tão abundante em cacáo, que toda he hum cacaol"*).[23] It seems likely that the Tikuna, or some other indigenous group, planted cacao on Aramaça and other islands along the Amazon.

It is possible that some ripe cacao pods floated down the Madeira, Purus, or Juruá and established the tree along the lower Amazon. More likely, though, most of the cacao on the Amazon floodplain has been planted, or is feral. In the middle of the nineteenth century, the English botanist Richard Spruce noted that the dense floodplain forest surrounding Lago dos Garças on the Amazon floodplain near Tupinambarana Island consisted mainly of wild cacao and urucuri palm.[24] The fact that the forest was densely packed with these economic species suggests that the cacao may not have been "natural." Indigenous people probably planted the shade-tolerant tree on parts of the Amazon floodplain.

The American oil palm was introduced to the Amazon floodplain long ago. Native to Central America, the moisture-loving palm was brought to various parts of the middle and upper Amazon before the arrival of Europeans and is often strongly associated with anthrosols. The prostrate palm is considered an indicator species for anthropogenic black earth.

Indigenous people also altered the vegetation on uplands adjacent to the floodplain. Ancient Brazil nut trees are common in the forest fringe surrounding villages on uplands bordering the middle Amazon floodplain, such as between Óbidos and Alenquer (figure 2.12). Locals recognize that the trees were planted; some of them may be descendants of seedlings planted, or at least encouraged, by the dense indigenous populations in precontact times. Human-induced concentrations of Brazil nut trees can be found along other stretches of the Amazon, such as the upland bluff 4 kilometers west of Itacoatiara, Amazonas. For 220 kilometers along the highway from Manaus to Itacoatiara, hardly a Brazil nut tree is to be seen. But as one approaches an upland bluff overlooking a seasonally flooded meadow on the outskirts of Itacoatiara, the distinctive silhouettes of the noble trees serve as a reminder of the antiquity of people in the region.

Repeated cycles of clearing and burning forest would have shifted the composition of the vegetation to species that tolerate frequent disturbance. Hunters and gatherers probably encouraged tucumã palm, especially around Monte Alegre. The abundance of curuá palm on upland sites near the Amazon floodplain in the vicinity of Santarém may be traced to frequent forest clearing in the past. Today, the palm is found only on disturbed sites. Forest disturbance may also have favored concentrations of some timber species. Along the lower Rio Preto in the estuary of the Amazon, for example, fires during an exceptionally dry year in the early 1900s may be partly responsible for the high density of virola. In upland areas,

Figure 2.12. Home gardens grading into forest enriched with Brazil nut trees along the floodplain margin. Several dead Brazil nut trees are in a field now abandoned to second growth dominated by babaçu palms. The skeleton Brazil nut trees, with their tall, straight boles, undoubtedly succumbed to fire set to clear the debris before planting crops. Near Alenquer, Pará, May 1996.

one of the most valuable timber species—mahogany—requires complete clearing in order to reproduce.

Pockets of savanna-woodland on uplands adjacent to the Amazon floodplain in the vicinity of Santarém are further testament to landscapes that have been shaped by human agency. During Pleistocene dry periods, when ice sheets covered the colder regions of the world, corridors of savanna-woodland most likely penetrated Amazonia. Indigenous people kept some of the savanna islands open by setting periodic fires; the forest has been advancing on the savannas ever since the aboriginal population crashed.

In the southern part of the Rupununi savanna in Roraima, the advance of forest is attributed to abandonment of agriculture and the reduction of man-made fires. In the Santarém area, farmers are virtually unanimous in asserting that the forest is closing in on the savannas. Farmers in Arapixuna, near the confluence of the Amazon and Tapajós, assert that the forest is rapidly taking over the grassland inland from their village, an observation echoed by residents of Vila Socorro on the shores of Lago Grande da Franca upstream from Santarém. This pattern of forest encroachment on savannas is also occurring in other parts of Amazonia, including the sinuous boundary between forest and savanna in the southern part of the region.

In addition to the extensive savannas on the interfluve between the Arapiuns and Amazon, anomalous scrub-grasslands are encountered on the outskirts of Santarém before ascending the upland scarp, and sporadically between Alenquer and Monte Alegre. Santarém is sprawling onto the partly anthropogenic scrub savanna, which contained sufficient quantities of feral cashew to support a juice and liquor factory in the middle of the last century. Indigenous groups probably introduced cashew to the Amazon long ago from its native area along the sand dunes of northeastern Brazil. In describing the scrub grassland in the vicinity of Santarém in 1851, Henry Walter Bates, one of the great nineteenth-century English naturalists, noted:

> The caju [cashew] is very abundant; indeed, some parts of the district might be called orchards of this tree. . . . It ripens in January, and the poorer classes of Santarem then resort to the campos and gather immense quantities, to make a drink or "wine" as it is called, which is considered a remedy in certain cutaneous disorders. The kernels are roasted and eaten.[25]

The scrub woodland on the upland bluff overlooking Paracari Lake, which is fed by a northern arm of the Amazon River as it bends south towards Santarém, also contains feral cashew and the sandpaper tree. The sandpaper tree, with its characteristic coarse leaves, is more typical of the *campo-cerrado* (scrub savanna) of central Brazil. According to locals, the grassy scrub at Paracari is also losing ground to forest. Herbert Smith, who spent 2 years in the Santarém area over a century ago, captures the ambiance of the scrub savanna at Paracari:

> Wandering over the sandy campos, we almost forget that we are in the Amazonas valley. Here the trees are scattered thickly over the surface, or gathered in little clumps, with bushes about their roots. They are low and spreading and crooked; rough-barked for the most part, and blackened by the yearly fires of herdsmen. . . . The landscape always reminds me of an old neglected orchard, where the trees have been left unpruned for years, and bushes and weeds have sprung up about their roots.[26]

Some scrublands, such as those on rocky outcrops, would probably survive indefinitely in the humid climate that returned after the close of the Pleistocene era, even without human-set fires, but people have clearly altered both the extent and species composition of relic savannas in Amazonia. Although the savannas of Amazonia are sometimes considered "natural," some people assert that they are man-made. Both positions are correct. In some situations, such as rocky outcrops or iron-ore hardened surfaces, scrubby woodlands are "natural." But the current extent and species contribution of most savannas in Amazonia are a result of human agency.

When indigenous people burned their fields during the dry season, fires likely escaped into the savannas and helped maintain their open nature even with the humid conditions that have prevailed since the end of the Pleistocene. Fires set on the savannas themselves can penetrate even into the surrounding forest during exceptionally dry years. This process was noted in 1899 by Paul Le Cointe, a French agronomist, when the dry season was so severe in the Óbidos area that fires devoured the prairies and consumed part of the forest. The smoke was so thick, he said, that navigation on the Amazon was unsafe at night.[27] During the exceptionally dry summer of 1992, farmers in the village of Arapixuna reported escaped fires from fields spreading onto the campos and then into older second growth at the edges of the grassland. In that year, the savanna scrub at km 20 of the Santarém-Alter do Chão road also caught fire.

Although lightning can cause fires in savannas, most of them have been set by humans. In postcontact times, livestock owners have deliberately burned some of the savanna pockets in order to encourage more nutritious forage for their cattle. One farmer in Vila Socorro suggested that the savannas are shrinking because they have been overgrazed; there is not enough combustible material to kill tree seedlings. Fires, at least ones hot enough to sear colonizing forest trees, have probably been more sporadic since the indigenous population crashed in the sixteenth and seventeenth centuries. The savannas of the Gran Pajonal in the Peruvian Amazon are regarded as a subclimax vegetation induced by human-set fires, and the islands of scrub-grassland in the Santarém area are also partly anthropogenic.

The Post-Contact Population Crash

As in many other parts of Latin America, the indigenous population plummeted in Amazonia after contact with Europeans. Introduced diseases such as smallpox, tuberculosis, and influenza spread quickly after contact with Orellana's expedition in 1542. Missionaries and explorers followed soon after, bringing with them a host of pathogens for which the indigenous populations had no resistance. In some parts of Latin America and the Caribbean, as many as 90 percent of the Indians died within decades after the Spanish and Portuguese arrived, and a similar fate befell many of the indigenous groups along the Amazon.

The proportion of the indigenous population that perished along the Amazon during the first hundred years after contact with Iberians is unknown, but it must have been substantial. In 1689, for example, Friar Samuel Fritz saw no people or settlements for six days below the confluence of the Urubu and Amazon rivers, a stretch of the Amazon where the Orellana expedition had to constantly fend off attackers scarcely 150 years earlier.[28] Massive destruction of indigenous societies continued throughout the colonial period with the added burden of slave raids, and along some of the tributaries of the Amazon, outright slaughter of Indian villages. In 1664, for example, the Governor of Pará ordered the burning of 300 indigenous long-houses (*malocas*) along the Urubu, a northern tributary of the middle Amazon.[29] The Urubu was aptly named: Vulture River. After the attack of 1664, the population of the once–densely settled Urubu plummeted, and even today, the river is virtually devoid of inhabitants. On a 2-week boat trip from the lower Madeira up the Amazon to Coarí in 1773, Francisco Sampaio lamented the complete lack of river traffic, or even a single house, along the 600-kilometer stretch of the once populous river.[30] Millions of indigenous people belonging to different cultural traditions and linguistic families had perished along the Amazon by the close of the seventeenth century.

While the indigenous population thinned dramatically, few immigrants arrived to take their place. Of the 5 million Africans brought to Brazil until slavery was abolished in 1888, only a relative handful came to Amazonia. At most, 50,000 slaves were brought to the Amazon, but the real figure was more likely closer to 30,000. Few Portuguese merchants in the relative outback of Amazonia could afford the high prices that slaves commanded. Most of the slaves were employed on sugarcane plantations, and to a lesser extent on cacao estates. Portuguese immigration was limited in the North to a few families engaged in commerce and the production of sugarcane alcohol.

Such a precipitous decline in the human population had major environmental implications for Amazonia. First, the forest enveloped abandoned villages and fields, both on the Amazon floodplain and adjacent uplands. Old second growth and managed fallows eventually returned to mature forest. Forest encroached on savanna islands. Second, many unique varieties of crop plants honed by generations of farmers with differing cultural tastes disappeared. With no one left to plant tubers, stems, or seeds, the painstakingly selected varieties were lost. It is possible that some plants that were once cultivated for food or other purposes also vanished, or reverted back to the wild. An immense stock of indigenous knowledge was lost.

Landscape Dynamics during the Colonial Period

Forest extraction and exploitation of turtle oil dominated the regional economy during the colonial period and only limited areas of the Amazon floodplain were cleared for agriculture. As the indigenous population declined after Orellana's expedition in 1542, the Amazon floodplain became progressively more empty of human life. By the mid-seventeenth century, little clearing was undertaken on the floodplain as most of the population had vanished. From the mid-1600s to the mid-1700s, the Amazon floodplain was a virtual demographic void, with few crops grown. Cacao was planted in Pará as early as 1678, but sizable groves of the cash crop were not established until the following century.

The planting tempo picked up again in the mid-eighteenth century, but not for food crops. Under edict of the Portuguese crown, cacao planting was encouraged on the higher banks of the Amazon floodplain from roughly Parintins (formerly Villa Nova and Villa Bella) downstream to the estuary. One Royal cacao grove on the right bank of the Amazon a little upstream from Santarém contained some 40,000 cacao trees in the middle of the eighteenth century. Although human population density remained low, sizable tracts of floodplain forest of the middle and lower Amazon were cleared for about a century to plant cacao. During the colonial era, cacao beans were exported to Europe, where chocolate was consumed as a beverage rather than eaten. The custom of drinking chocolate, which predates coffee by more than a century in Europe, was always considered a luxury.

"Wild" cacao was also collected from upland forests in Amazonia, although it is probably truly native only west of the Madeira. Cacao was treated mainly as an extractive enterprise, since little care was expended on the plantations along the Amazon. Rather, Indian slave labor was dispatched to collect the pods and prepare the beans for market. Some of the extensive plantings of cacao along the middle and lower Amazon during the colonial period were eventually abandoned and became feral. A large cacao plantation on Careiro Island near Manaus, for example, was being engulfed by forest by the middle of the last century. People still harvest wild descendants of these plantings in the Amazon estuary and on islands near Cametá on the lower Tocantins.

Cacao plantations were especially common along the Amazon River and its side-channels between Parintins and Monte Alegre by the middle of the nineteenth century. In the nineteenth century, many travelers described the banks of the Amazon between Juriti and Alenquer as being continuously lined with cacao orchards varying from 50 to 200 meters wide. A little further downstream near Prainha, the landscape of the Amazon in 1873 was described thus by Brown and Lidstone:

> The narrow bands of cacoa trees bordering the edges of the Lower Amazon form one of the most marked and characteristic features of its scenery. Their light-green leaves, tinged with brown, contrast in a pleasing manner with the tall dark wall of forest foliage behind.[31]

Much of the cacao was in small holdings, but some estates along the banks of the middle Amazon contained some 30,000 cacao trees. Not surprisingly, cacao was the single most important export from towns along the middle Amazon, such as Itacoatiara and Parintins, in the 1890s. Downstream from Monte Alegre, however, cacao plantations were less common, confined mostly to the south bank of the Amazon. Today, cacao groves and sheds with drawers for driving the beans are a rare sight (figures 2.13, 2.14).

Above the confluence with the Madeira, clearing on the Amazon floodplain for cacao was generally modest. In 1850, for example, Bates, who lived and traveled along the Upper Amazon for more than four years, surmised that the total cultivated area along the Amazon from its juncture with the Negro to the Andes amounted to only "a few score acres."[32]

The only other crop planted on any scale on the Amazon floodplain during the colonial period was sugarcane, grown mostly for the manufacture of white rum (*aguardiente* or *cachaça*) rather than for sugar. Some of the cachaça was exported to Portugal, along with small quantities of anisette using aniseed imported

Figure 2.13. A cacao orchard in cleared forest on the floodplain near Santarém, Pará, in the mid-1870s. From Smith, H. H., *Brazil: The Amazons and the Coast* (Charles Scribner's Sons, New York, 1879), p. 260.

from Portugal or Gibraltar. Sugarcane was not grown on a large scale in the Amazon because it was not allowed by colonial authorities to compete in world markets with sugar produced in Bahia.

Most of the sugarcane cultivated in the Brazilian Amazon during the colonial period was confined to the estuarine area of the Amazon. In 1751, sixty-six sugarcane mills were operating in Pará, forty-two of them engaged exclusively in the manufacture of "firewater." The sugarcane mills, scattered at various locations near the mouth of the Amazon, harnessed tidal power to squeeze the cane. Several missionary orders were involved in establishing sugarcane plantings in the estuary of the Amazon, including the Carmelites, who even had drainage canals dug on parts of the floodplain, such as at Fazenda Murutucú on the outskirts of Belém, to facilitate cultivation of the crop. Because the population was so sparse, only minuscule areas of forest were cleared for food production on either uplands or river floodplains. Furthermore, the urban population, such as it was in the seventeenth, eighteenth, and nineteenth centuries, depended heavily on food imports from other regions, including Europe.

In spite of a mini-boom in cacao and sugarcane in the Amazon estuary, the region's population remained sparse until well into the twentieth century. In 1800, only 90,000 people were recorded in Amazonia, and by 1840, the population still had not surpassed 130,000. The paucity of settlements that prevailed for much of the colonial period was a far cry from the days when flotillas of war canoes launched out onto the Amazon from villages that whitened the horizon with their sun-bleached thatch roofs. Drums that once summoned indigenous groups for festivals or sounded the alarm had fallen silent.

With two cities in the Brazilian Amazon currently surpassing the million mark, it is hard to conceive that they were little more than overgrown villages or small towns until the twentieth century. Belém, founded in 1616 at the mouth of the Amazon, had only 200 inhabitants in 1637, increasing to 5,000 residents by 1718. In 1832, the city boasted 12,467 inhabitants. By 1848, the population of Belém had grown only to 15,000, depressed in part by the peasant revolt (*cabanagem*) of the mid-1830s. Propelled by the rubber boom, Belém's population leaped to 100,000 by 1894. By the year 2000, Belém is likely to have nearly 2 million inhabitants.

The region's second largest city, Manaus, had only 6,000 inhabitants in 1867, rising to a modest 20,000 by 1893. In the late 1990s, the population of Manaus had surpassed 1.5 million. Santarém, the third largest city, had only 2,000 inhabitants in 1849 (figure 2.15), but by 1994, the population had swollen to 300,000 (figure 2.16). The spurt of urban growth throughout the Amazon in the last few

Figure 2.14. Laborers drying cacao on the floodplain near Santarém, Pará, in the mid-1870s. The person in the left foreground is splitting open the pods to obtain the cacao beans. From Smith, H. H., *Brazil: The Amazons and the Coast* (Charles Scribner's Sons, New York, 1879), p. 113.

Figure 2.15. Santarém at the confluence of the Tapajós and Amazon rivers in 1849, when the population was 2,000. From Spruce, R., *Notes of a botanist on the Amazon and Andes*, vol. 1 (Macmillan, London, 1908), p. 64.

Figure 2.16. Santarém in June 1994, when the population was approximately 300,000 residents. The main church, Nossa Senhora da Conceição, still stands, but businesses and apartments have replaced the thatched huts. Sailing ships no longer grace the waterfront and a couple of floating diesel-fuel stations can be seen to the right of the church.

decades has created a strong demand for beef and other agricultural and extractive products, and these driving forces are transforming the landscape along many parts of the Amazon floodplain.

In 1867, 110 white southerners from the United States settled in the Santarém area in order to flee the changing social and political scene back home following the abolition of slavery. Land was relatively cheap in the Amazon and slavery was permitted in Brazil for twenty-three years after the U.S. civil war. At least 20,000 *confederados* eventually migrated to Brazil, with the largest contingent settling in and around Americana in the southern state of São Paulo. Few of the *confederados* who chose Santarém in the Amazon had any experience with farming or even physical labor. Locals complained that the confederates were disorderly and drunk much of the time. Most of the confederates left within a few years, and those that remained tilled the uplands, rather than the floodplain of the Amazon.

One of the most successful *confederado* properties, the Taperinha plantation owned by Mr. R. J. Rhome in partnership with the Baron of Santarém, bordered the Amazon floodplain downstream from Santarém. Site of a shell mound with 8,000-year-old pottery, the Taperinha plantation used 40 African slaves to cultivate sugarcane, and to a lesser extent tobacco, on the upland bluffs overlooking the Maicá River, part of the Amazon floodplain. By the 1870s, much of the landscape around the plantation house was denuded to produce sugar, molasses, and especially white rum (figure 2.17). Today, the uplands overlooking the plantation house are largely forested again (figure 2.18)

A first-time visitor to Taperinha might assume that the lush forest carpeting the scarp slopes and plateau is pristine. But Taperinha serves as a good example of the many human-induced transformations of the Amazon floodplain and adjacent uplands. One can imagine hunters and gatherers foraging along the gentle footslopes of the Taperinha bluffs, exploiting plant and animal resources from both the floodplain and upland. From roughly 8,000 years to the time of contact, the upland and *várzea* forests of Taperinha may have ebbed and flowed several

Figure 2.17. The Taperinha plantation near Santarém in the mid-1870s. The hills have been cleared to plant sugarcane and tobacco. Cut cane was tossed down the chute leading from the crest of the scarp to the plantation house. From Smith, H. H., *Brazil: The Amazons and the Coast* (Charles Scribner's Sons, New York, 1879), p. 152.

Figure 2.18. The Taperinha property in March 1993. The hills have largely reverted to forest, and the main activity now is grazing cattle and water buffalo at low water on the floodplain. The ranch house in the center is approximately two hundred years old. The large tower on the right-hand corner of the house is a later addition. Remnants of dikes to convey water from a stream can be seen immediately behind the buriti palms near the house.

times in response to changes in human population density. The area was certainly densely settled in the past, judging by the extensive anthropogenic black earth on the upland bluffs that were responsible for the good sugarcane yields during the colonial period. Sugarcane was probably first planted around Taperinha in the eighteenth century by a previous Portuguese owner; at that time, the area was largely forested because the indigenous population had died off or had been captured for the slave trade. In the early part of the twentieth century, the sugarcane plantation was abandoned, and the forest reasserted itself.

Lifeways and landscapes have changed dramatically along the Amazon since the middle of the nineteenth century. Difficulties in securing labor for a zoological expedition from Harvard University at Tefé in the last century highlight how the natural resource base and technologies have changed:

> We had at first some difficulty in finding servants; at this fishing season, when the men are going off to dry and salt fish, and when the season for hunting turtle-eggs and making turtle-butter is coming on, the town is almost deserted by the men.[33]

Today, the giant Amazon river turtle, which once furnished hundreds of millions of eggs yearly for the oil trade, is all but gone. While rural people still salt and sun dry fish, temporary camps set up along margins of lakes to preserve fish, especially pirarucu, are no more. Most of the fish catch is currently put on ice by commercial fleets that range hundreds of kilometers from port. Ice only became available on a significant scale in the early 1900s, and since then fishing fleets have exerted enormous pressure on many of the formerly highly productive fisheries. The steamships that once plied the Amazon, using large amounts of driftwood for fuel, have been replaced since the 1930s by boats powered by diesel engines.

The Rubber Boom and Landscape Changes in the Twentieth Century

The rubber boom, the most famous or infamous of the extractive products of Amazonia, had little immediate impact on the landscapes of Amazonia outside of urban areas. Rubber dominated the regional economy for only a few decades, but it had a lasting impact on the culture of the region. The rubber boom started in earnest in the 1870s, but was essentially over by 1920, when rubber estates in Southeast Asia came on line and undercut the world price for the commodity. Over 300,000 people migrated from the thorn-scrub backlands of the Northeast of Brazil to tap rubber in the Amazon between 1872 and 1910, a major cultural influx considering the total population of the Brazilian Amazon in 1870 was only 323,000. The cultural imprint of drought refugees from the Northeast is still felt in many parts of the Brazilian Amazon, ranging from the legacy of crops and introduced varieties to livestock, music, and dance.

The immediate impact of rubber tappers on the landscapes of the Amazon River itself was negligible. Most rubber tappers headed for tributaries of the Amazon on steamships, especially to the headwaters of the Madeira, Purus, Juruá, and to a lesser extent the Xingu and Tapajós. The Amazon River served as a conduit for the settlers, not a place for work or settlement. The rubber trees found on the Amazon floodplain today appear to have been largely planted. If any rubber trees grew naturally on the Amazon floodplain in the latter part of the nineteenth century, they were not dense enough to be of any interest to tappers except in a few locations, such as on some islands at the mouth of the Amazon.

Little deforestation took place even in headwaters during the rubber boom because the tappers did little farming. They were grubstaked by patrons who provided them with basic foodstuffs and other materials necessary for their daily task of tapping trees spread out in the forest. In return, the rubber tappers handed over their smoked latex balls to their bosses during the four- to six-month collecting season. A few trees were cut down to build huts, but otherwise the tappers had little impact on the forest cover. Rubber is tapped during the dry months; when the rubber season was over, no time was left to cut down the forest and let it dry for a good burn to plant crops. Few tappers even had much time to hunt. In the Peruvian Amazon, some tappers maintained small subsistence plots, but these were mostly indigenous people only partially engaged in the rubber trade. Overall, deforestation rates were much lower than in precontact times.

If anything, the influx of rubber tappers reduced deforestation rates because they sometimes fought with surviving indigenous groups and annihilated hostile villages. When rubber tappers established amicable relations with Indians they probably contaminated many of them with introduced diseases that subsequently decimated villages. The further reduction of indigenous people along the tributaries of the Amazon and in the headwaters meant fewer fields were being cleared in the forest to grow crops. The increased economic activity associated with the rubber boom did not spur an overall increase in crop planting; rice and cacao production declined, while the area devoted to manioc, tobacco, and sugarcane increased modestly in some cases. Labor formerly devoted to such crops as cacao on the Amazon floodplain was siphoned off to tap rubber. Cacao towns, such as Alenquer and Monte Alegre, became depressed communities during the rubber boom.

After the rubber season was over, tappers generally migrated to urban areas in search of employment, or if they were lucky, lived off their earnings from the sale of rubber. Only after the collapse of the rubber boom did some of the rubber tappers, by then often married and with families, turn to farming. Most of the current farmers on the Amazon floodplain within Brazil trace their origins to the Northeast, particularly the drought-prone state of Ceará. They often married local women of indigenous descent. After about 1910, then, *Nordestinos* and their families began clearing the forest in earnest to grow crops along the Amazon and some of its tributaries.

The cultural impact of Northeasterners along the Amazon has been marked because they occupied a nearly empty land. *Nordestinos* have left their cultural imprint by introducing the carnaúba palm, bringing in more goats, and introducing the *bumba-meu-boi* dance. Carnaúba palm is native to the Northeast of Brazil, where the fronds are harvested to extract a hard wax for a variety of uses, including car wax. Carnaúba palm is occasionally encountered on the Amazon floodplain, such as along Igarapé do Mamauru near the village of Flexal in the Municipality of Óbidos. Carnaúba palm was brought as a reminder of home for many of the settlers on the Amazon floodplain and adjacent uplands, rather than for commercial purposes. Goats were present in small numbers in the Amazon before the rubber boom, but their numbers rose significantly after 1910. Goats are well adapted to the harsh, dry environment in the interior of the Northeast, and *Nordestinos* undoubtedly fetched some of the diverse, mixed breeds for their farms on the Amazon floodplain.

Bumba-meu-boi or *boi-bumba* is another cultural fixture of life and landscapes along the Amazon. All sizable villages and towns along the Amazon celebrate the *boi* (cow), a festivity that arose in the northeast of Brazil and traces its origins to a blend of African, Iberian, and indigenous traditions. In June and July, processions and dances are organized around the theme of a cow that is eventually killed after a ritual dance. At Arapixuna near Santarém, districts of the village strung out along the upland bluff overlooking the Amazon floodplain hold competitions during the Círio Catholic festival in late July to see who has come up with the most animated costume of the cow. In 1996, only three districts entered costume cows in the competition, and old-timers griped that every year enthusiasm for the festival seems to be waning.

In other parts of the Brazilian Amazon, however, boi-bumba is taking off. At Parintins, an Amazon riverside town by the border between the Brazilian states of Amazonas and Pará, the boi festival has emerged as a big business. Townsfolk organize a major festival around boi-bumba every June that attracts tourists from as far away as southern Brazil. VARIG, (Viação Aerea Rio Grandense), Brazil's largest airline, featured the festival on the cover of the September 1997 edition of its international flight magazine. Boi has caught on as the hot new dance in Manaus and Belém, and may yet make its way to New York and Paris as did the lambada in the early 1990s. The prominent role played by a cow in this festival underscores the importance of cattle ranching along the Amazon, a major agent in landscape transformation.

The Modern Period

The population of Amazonia is only now reaching the levels attained before the arrival of Europeans. The major difference is that the population is predominately urban. The countryside, although slowly filling with people since the onset of the rubber boom, is far less densely settled than it was in 1500. Towns and cities have acted like magnets in the last few decades, attracting people from the Amazon floodplain and adjacent uplands.

With a relatively low population density in rural areas, and the fact that development planners have largely ignored the Amazon floodplain, one might expect that the forest and other habitats important for wildlife would be in relatively good shape. Most of the development efforts in Amazonia have focused on interfluvial areas, characterized by ambitious colonization schemes, large ranching and plantation operations, and mega-mining projects. Although the Amazon River itself has been spared from often-controversial development efforts, cultural and economic changes are nevertheless inflicting major environmental damage. The growing urban markets in Amazonia present a dilemma for people and the environment on the Amazon floodplain. On the one hand, they provide new opportunities for farmers to prosper and experiment with novel crops. On the other hand, they create huge markets for beef, thereby accelerating the stampede towards cattle and water buffalo ranching on the floodplain.

A Forest Cornucopia

Floodplain forests are biologically diverse, in spite of the annual floods, and people have learned to tap this biodiversity for food, medicines, and other uses. Inhabitants of the Amazon floodplain gather fruits, nuts, barks, and other plant products from a wide variety of habitats, ranging from flooded forests to floating meadows and weedy communities around garden plots and homes. The composition of the *várzea* forest changes markedly along the Amazon. Some species are found only in the middle or upper stretches, while other plants are widespread but occur in concentrations. Gathering activities follow a seasonal rhythm. Some species fruit during the dry season, while others release their fruits at high water to be dispersed by currents or fish. Floodplain residents thus benefit from a year-round supply of food, building materials, and other supplies for the household gathered from the wild.

Although extractive products are an important supplement to diet and income on the floodplain, the gathering of wild plant products alone does not normally provide sufficient income to justify "saving" all of the forest. Intensified use of forest products should be seen as part of an interlocking system of land use, rather than a "sustainable" approach to development. If the biological and economic value of forests is better appreciated, however, they are more likely to survive.

In addition to a smorgasbord of fruits, nuts, and other edible plant stuffs, floodplain forests also contain gene pools of economically important plants that could be tapped to promote more biologically diverse and productive agriculture, particularly agroforestry. The Amazon floodplain contains wild or feral populations of several commercially valuable crops, such as cacao and rubber, and these genetic reservoirs could eventually be tapped by breeders for useful breeds—if their habitats are left standing. Some of these genetic reservoirs have already been used to upgrade existing crops. The American oil palm, for example, grows wild along certain stretches of the middle and upper Amazon and has been crossed with its cultivated cousin, the African oil palm, to improve the quality of the oil, impart disease resistance, and to shorten the palm for ease of harvesting.

Any one of the numerous wild plants of the Amazon floodplain that are currently gathered for use in the kitchen or in cottage industries could emerge as tomorrow's "hot" crop, as consumers in Brazil and abroad are keen to try new tastes in their cuisine. The spectacular emergence of the kiwi fruit from its obscure origins as a minor crop in southern China, for example, indicates that there is always room in the crowded produce sections of supermarkets for appealing and versatile new fruits and vegetables.

Many of the plant resources of the Amazon floodplain are poorly known or underutilized. Much potentially useful knowledge about the plants of flooded forests, floating meadows, and lakeshores has been lost as indigenous people have disappeared. According to a French agronomist, Paul Le Cointe, who lived at Óbidos in the early twentieth century, people employed more than seventy-five plants along the Amazon floodplain in the 1930s for a variety of purposes.[1] Some of those species are now difficult to find because of drastic habitat modification, or because manufactured products have replaced them. In spite of rapid ecological and cultural change, however, rural inhabitants along the Amazon still harvest numerous wild plants for a wide variety of purposes. The knowledge and expertise of farmers and fisherfolk re- garding the natural history of wild plants is a valuable resource that could be used to foster a more rational development of the region, yet such expertise is largely ignored.

A Forest Smorgasbord

Fruits, hearts-of-palm, legumes, and nuts collected in the seasonally flooded forest provide valuable vitamins and protein to the diet. Fruits take the place of vegetables as a major source of vitamins in the daily cuisine of most rural peoples native to Amazonia. Some of this foodstuff also makes its way to local and regional markets, supplementing family incomes.

Palms, in particular, feature prominently in the diet of people along the banks and lake margins of the Amazon, among which are the açaí palm, the buriti palm, the caraná palm, and the tucumã palm. Other fruits much appreciated include the fruits of the yellow mombin and bacuri trees, as well as tree legumes such as ingá and marimari. People also avidly seek nuts in floodplain forests, especially those from the supucaia tree.

Açaí Palm

The grape-sized fruits of açaí palm are collected and mashed to provide a satisfying drink with the consistency of a milkshake. People have appreciated the distinctive flavor of açaí in Amazonia for millennia, as evidenced by the abundance of seeds in archaeological sites on Marajó Island at the mouth of the Amazon. Graceful açaí palms form dense groves in the Amazon estuary, where their purple fruits are still gathered in large quantities, especially for urban markets. In the

Amazon estuary, farmers frequently sow açaí seeds in garden plots after harvesting annual crops to promote regeneration of the lucrative palm. The fruits bring in significant income for riverine dwellers (*ribeirinhos*) on alluvial islands of the lower Amazon. On Combu Island near Belém, for example, Anthony Anderson and collaborators have found that family income exceeds $4,000 per annum, most of it derived from the sale of açaí fruits.[2]

Although açaí is native to parts of the Amazon floodplain, its current distribution and density is the result of generations of enrichment planting. Açaí occurs sporadically along the middle and upper stretches of the floodplain, and is less important in the local diet and economy above the confluence of the Amazon and the Xingu. An anomalous concentration of açaí is found in the backswamp forest of the floodplain in the vicinity of Murumuru near Santarém, where stands are sufficiently dense to support collecting of fruits for the local market (figure 3.1). Parts of the backswamp forest that have been cleared for short-cycle crops are sometimes planted to açaí after the harvest. Indigenous people probably introduced açaí to the Amazon floodplain at Murumuru from seedlings gathered in forested streams that drain the scarp of the Santarém plateau (*planalto*), or along the margins of springs at the base of the steep upland bluff. Açaí has also been introduced to the floodplain on Careiro Island near Manaus. Another species of açaí, *Euterpe precatoria*, normally found in upland forests, has long been planted on higher parts of the floodplain along the Upper Amazon.

Figure 3.1. Collecting fruits of açaí palm in a backswamp forest. The older boy has shimmied up several of the palms to cut the fruit stalks and has joined his brother in stripping the fruits off the stalks. It is late in the afternoon, and the fruits will soon be taken home; in the early hours of the morning a buyer will take the fruits on the two-hour bus trip to Santarém for sale that day. Murumuru near Santarém, Pará, June 1996.

A similar process of managed fallows and enrichment planting to encourage açaí occurs in the estuarine area of the Amazon. Along the Maracá River, an affluent on the north bank of the Amazon in Amapá, some farmers maintain nurseries of the palm for planting after annual crops and in light gaps created in the forest when some of the trees are removed. In some areas açaí stands become so dense due to human agency that they are locally recognized as a vegetation type: *açaízal*.

If the fruits of açaí are destined for market, they are harvested the day before for early-morning delivery to urban centers (figure 3.2). Açaí fruits spoil if they are kept at ambient temperatures for more than thirty-six hours or so after they are picked. Most of the fruit is gathered from areas where resource rights are generally respected. Farms are individually owned, even if official paperwork is wanting, and few disputes occur over land ownership. In riverine areas, properties of small farmers are typically narrow, with about one hundred meters of frontage along the riverbank and one to two kilometers back from it. Açaí within property boundaries is therefore considered privately owned. At Murumuru, an exclusive extraction zone of roughly two hundred meters into the floodplain forest is recognized from the boundary between uplands and the floodplain. For communities such as Murumuru, which is located along the divide between uplands and the floodplain, properties mostly extend inland. Within the 200-meter exclusive extraction zone, fruits of açaí and other plants are generally recognized as belonging to the owner of the adjacent upland property. Few cases of "stealing" of the fruit are reported because property owners build their houses close to the boundary between

Figure 3.2. Açaí fruits, gathered from floodplain forest in the Amazon estuary, being loaded on to a cart shortly after sunrise for delivery to açaí stores. Moju, Pará, August 1992.

terra firme and the floodplain and can hear people at work up to a couple of hundred meters within the swamp forest. The backswamp forest extends up to a kilometer into the Amazon floodplain, and açaí palms beyond the 200-meter exclusive collecting zone are an open to all community members and outsiders.

The task of shimmying up the slender trunks of açaí, which can reach almost straight up for twenty meters, is left to teenage boys and girls and young men. A small, pliable sling (*peconha*) is placed around the feet to serve as a brace for the climber. In the past, the climbing loop was typically made from young açaí fronds or the leaf base (*capota*), but today the sling is mostly fashioned from polypropylene sacking. Once at the top, climbers cut the fruit stalks with knives and drop or carry them to the ground. On the forest floor, the collectors and any waiting family members strip the fruits manually from the stalks onto a polypropylene sack or other material to keep them from rolling into the dirt; the fruits are then poured into 60-kilogram sacks or baskets and carried home. In some cases, the fruits are stripped directly into a basket made from a wide variety of wild plants, including cipó ambé, a Philodendron-like plant that produces pliable tap roots that extend down from the tree tops.

The fruits are transported to urban centers by boat, or from some communities such as Murumuru, by early morning bus. Farmers rarely hawk their fruit in marketplaces. They either sell their fruit to local buyers who take the fruit to market, or once in town, hand over the fruit to wholesalers (*atravessadores*). The largest market for açaí fruit is the *feira do açaí* (açaí market) adjacent to Belém's Ver-o-Pêso (see-the-weight) market, where dozens of wholesalers operate (figure 3.3). In the

Figure 3.3. A wholesaler (*atravessador*) in the *feira do açaí*, the Amazon's largest market for açaí fruit, in Belém, Pará. It is seven A.M., and the market has already been operating for a couple of hours; it will close within an hour until the next morning. September 1997.

smaller towns, a porter (*marreteiro*) is hired to push the fruits in a flatbed wheelbarrow or pull it in a cart to an açaí shop after the deal is closed. In cities, trucks or vans take the fruit to stores. In July 1996, farmers at Murumuru could sell açaí to local buyers for three to four dollars per eighteen liters, the measuring unit of a kerosene can. The buyers then take the fruits to town and sell them to wholesalers. If they are prepared to take the three-hour bus ride to Santarém, farmers can get an additional one dollar per eighteen liters for their trouble. Farmers generally opt to take the fruits themselves only if they have other business in town.

Prices are lowest during the main harvesting period from July to November, which coincides with the height of the dry season. Açaí also fruits during the wetter months, but yields are more modest and prices for fresh açaí juice rise sharply. Prices for açaí ice cream remain relatively constant, because it is made from frozen pulp, which can be stored year-round. In 1996, consumers in Santarém were paying two dollars per liter for açaí juice at the beginning of the main harvest (July), a considerable markup. In the lower Amazon, açaí prices are a little lower because of the greater abundance of the palm. In Breves, for example, customers were paying one dollar per liter for açaí juice toward the tail end of the season in November 1995.

Once the fruits have been sold in town, most of them end up in small açaí shops where they are spun in simple, "cottage-industry" blenders to extract the pulp (figure 3.4). Açaí shops are easily recognized because of their red flags displayed out front, a practice one might expect more from a butcher than a store specializing in a tropical fruit. Another telltale sign of an açaí store is the mound of seeds sometimes discarded in the street after the pulp has been removed. Not all açaí shops have associated piles of seed because the seeds are sought after by nurseries to make potting soil. For example, a coconut plantation near Mosqueiro, Pará, uses açaí seeds (*caroço de açaí*) to establish coconut seedlings in plastic bags. Similarly, farmers sometimes employ depulped açaí seeds in organic soil mixes for raised vegetable and spice beds.

Açaí juice is typically sold in plastic bags to be taken home and eaten fresh, particularly in the late afternoon, or is frozen to make açaí ice cream. Sometimes people partake of açaí right at the store, served in China or glass bowls or black calabash gourds (figure 3.5). Açaí juice is often sweetened with sugar and thickened with tapioca or manioc flour to form a porridge that is eaten with any meal, but especially for lunch or dinner. A bowl of thick açaí is so filling that for some it suffices for supper. Purists appreciate the savory flavor of açaí "straight." At home, açaí is eaten in ceramic bowls as a side dish, or for dessert. Açaí is thought to provide a good "balance" when highly salted foods are on the menu, such as salted and sun-dried shrimp.

Young and old congregate at açaí stores towards the close of day to chat between spoonfuls of açaí, which can be ordered in thick form at a higher price, or for the cost-conscious in a more diluted concentration. Açaí stores thus serve as informal meeting places to gossip and exchange information. In communities along the Amazon and the lower courses of its affluents in Pará and Amapá the day is not complete without a bowl or two of açaí.

Açaí contains traces of only a few vitamins, and only modest amounts of iron, calcium, and phosphorus. But the fruit is a good source of calories, especially when

Figure 3.4. Preparing açaí juice with a specially constructed blender. The fruits are poured into the top of the machine and water is added according to the desired thickness. The viscous juice collects at the bottom in an aluminum pan. Some açaí shops sell two grades of juice: a concentrated premium grade (usually called *especial*), and the more diluted regular kind. Mosqueiro, Pará, June 1994.

Figure 3.5. Açaí juice ready for consumption on the spot, or sold in a plastic bag to be taken home. The lady is wearing a cap with the logo of a soccer team, a sport for which Brazilians share a passion. Mosqueiro, Pará, June 1994.

Figure 3.6. Heart of açaí palm ready to be picked up by an approaching boat shortly after sunrise. The heart-of-palm was harvested from a cultural forest behind the man's home the previous afternoon. The product will most likely be consumed in France. Comunidade São João, lower Maracá River, Amapá, May 1996.

taken with sugar and tapioca or manioc flour. It is the unusual, savory taste of açaí, rather than its nutritional qualities, that drives the market for this fruit. Inhabitants of Pará are so enamored with it that they have an expression extolling its seductive properties: *"Chegou no Pará, parou; tomou açaí, ficou"* ("Came to Pará, and stopped; drank açaí, and stayed"). Versions of this proverb have been around for more than a century. In 1865, for example, the Louis Agassiz expedition to the Amazon, was told in Belém: "Who visits Pará is glad to stay, who drinks assai [açaí] goes never away."[3] Although açaí can be an acquired taste, Pinduca, a well-known regional singer, refers to the enchanting qualities of the fruit in one of his carimbó songs titled "Quem vai ao Pará, parou." Carimbó is a samba-like music tradition from northeastern Pará often flavored by references to regional products.

Açaí is also harvested for its heart-of-palm (*palmito*), primarily for the numerous canneries in the estuarine area. Açaí palmito is exported and consumed by the well-to-do in urban areas of Brazil. By the early 1990s, palmito canning in the Amazon estuary was generating several million dollars a year in sales. Along the lower Maracá River in Amapá, an affluent of the Amazon, riverine dwellers harvest palmito in the afternoon, stack the baseball bat-sized palm hearts on elevated platforms by the "port," and sell them to itinerant buyers the next morning (figure 3.6). Heart-of-palm is harvested by felling the trees close to the ground and cutting out the young "spike" of developing leaves. Older trees are usually targeted for harvesting because they are taller and thus more difficult for fruit gatherers. Although harvesting is a year-round activity, it picks up during the rainy season when fruit yields drop.

If managed rationally, açaí groves can provide palmito on a sustained basis because the palm sprouts again when lopped off near the ground. Selective pruning for heart-of-palm enhances fruit production. On the other hand, the palm dies if the shoot is cut out at the top and the tree remains standing. Some clandestine operators shimmy up açaí to extract the palmito, thereby destroying the resource. In some areas, conflicts have arisen between those who depend on açaí for fruits, and those who harvest palmito. For the most part, though, açaí is a controlled-access resource, and therefore individual landowners make decisions about how they want to manage their stands. In 1997, açaí fruit prices in northern Marajó and other areas of the Amazon estuary were sufficiently strong to discourage excessive harvesting of groves for palmito.

Buriti Palm

The widespread buriti palm (also called *miriti* in Brazil and *aguaje* in Peru) is especially concentrated along the upper Amazon where large quantities of the pulp from its vitamin C–rich fruits are marketed in Iquitos. The fruits of this fan-leafed palm contain three times as much vitamin A as do carrots. Buriti is also prolific: In a survey of two million hectares of forest near Iquitos, 80 percent of the trees were buriti palm. A single specimen alone can yield 5,000 golf-ball–sized fruits annually (figure 3.7). Buriti stands are so dense along the Ucayali

Figure 3.7. Buriti fruits gathered from the floor of a floodplain forest near Afuá on Marajó Island, Pará, September 1997. The fruits will be eaten as a snack.

that they form a distinct vegetation type: *aguajal*. Buriti is also found in relatively dense stands along clear- and even black-water rivers, as well as on poorly drained savannas. The concentrations of buriti palm in lakes of the upper Amazon floodplain may be due in part to artificial enrichment by the once dense indigenous population. In Brazil, a grove of buriti—whether "natural" or manmade—is called a *buritizal*.

Deep-orange buriti juice and ice cream are regional delicacies from Pucallpa on the Ucayali to Belém at the mouth of the Amazon. The scaly, red skin of the fruit is peeled, and after the pulp is removed from the seeds, it is placed in plastic bags and sold fresh in local markets. The main fruiting season for buriti along the Amazon in Brazil stretches from March to December.

A mature buriti palm towers up to 25 meters, and its smooth, straight stem makes for hazardous climbing. The fruits are usually gathered as they fall into the water or onto muddy ground. In the Peruvian Amazon and in the Rio Negro watershed, the shortsighted practice of cutting down the female trees to obtain the fruits has led to local scarcity of the once abundant product. Yet felled buriti palms are not completely wasted: Locals return later to feast on the larvae of the palm beetle, which they pry loose from the rotting trunks. The grubs of this beetle, known as *suri* in Peru, thrive on other species of dead palm and are rich in protein. Dead, standing buriti palms are favorite nesting sites for blue and yellow macaws, so continued destruction of buriti stands will likely reduce numbers of the handsome megaparrot. Fortunately, this destructive method of harvesting does not appear to be common along the middle and lower Amazon.

In the lower Amazon, buriti palms are sometimes felled for a variety of other reasons. In the vicinity of Afuá on the northwest coast of Marajó Island, farmers sometimes collect the sugary sap of buriti by felling a mature tree and scooping out a hole in the trunk. The hole is then covered with leaves to deflect rain. After the sap, called *mel de buriti* (buriti "honey"), has seeped into the hole for a day or two, it can be scooped up and taken home to drink. A mature buriti is sometimes felled to provide a convenient ramp for foot traffic at the "port" of small farms. And loggers operating near Afuá use buriti trees to float hardwoods, such as heavy and dense anani, so that they can be towed to a sawmill.

Caraná Palm

A shorter cousin to buriti, the caraná palm is confined to clear- and dark-water backswamps. Along the Amazon, it is typically found in almost permanently flooded forest close to the interface between the floodplain and adjacent uplands, such as in the vicinity of Murumuru. Unlike buriti palm's smooth trunk, the stem of the caraná is studded with thick, menacing spines. Consequently, its fruits are harvested during the wet season by tying a hooked knife (*foice*) to a pole long enough to reach the fruit clusters about 8 meters above ground. Pulp from caraná is mixed with water and sugar to make a welcome drink. Neither the fruits nor pulp of caraná seem to make their way to markets; it is one of those "minor" fruits, esteemed locally but not captured in any statistical yearbooks or market surveys.

Tucumã Palm

The fruits of two species of tucumã palm are eaten in the Brazilian Amazon. One, *Astrocaryum vulgare*, is mostly associated with disturbed sites and grows on both uplands and higher parts of the Amazon floodplain in Pará and Amapá. It typically occurs in small clumps and both its distribution and density have much to do with human agency. The bright orange fruits of *Astrocaryum vulgare* are rarely encountered in markets, but are nevertheless relished by inhabitants of the Amazon floodplain as well as upland sites. The other species, *Astrocaryum aculeatum*, is strictly an upland palm and is more common in the states of Amazonas and Acre, although it occurs sporadically in Pará. Also called tucumã, *Astrocaryum aculeatum* produces larger fruits and generally grows singly rather than in clumped groups. Its green-skinned fruits find a ready market in towns of Amazonas, particularly Manaus.

Just west of the Camará River on Marajó Island, hundreds of tucumã (*A. vulgare*) are found on slightly elevated portions of the seasonally-flooded savanna. This island was once densely settled by indigenous groups, and it seems likely that the extensive groves of tucumã are vestiges of artificial enrichment; indeed, tucumã seeds have been found in archaeological mounds on Marajó Island. Further upstream near Monte Alegre, seeds of tucumã have been located in archaeological sites overlooking the Amazon floodplain that were occupied by hunters and gatherers towards the end of the last Ice Age.

The common or "vulgar" species of tucumã is abundant in second growth on upland bluffs overlooking the floodplain in the vicinity of Santarém, especially between Arapixuna and Aninduba. Inhabitants of the floodplain, particularly children, avidly gather the mealy and oily fruits of *Astrocaryum vulgare*. The fruits are also cut down with a knife tied to a pole; long, down-pointing spines occurring in whorls along the trunk are an effective deterrent to climbers. The fruits have evolved their conspicuous appearance to attract seed dispersal agents, and their color signals a high vitamin A content, which is three times greater than the carrot. The somewhat greasy texture of the fruit also signals its high fat content. While fats are almost taboo in relatively well-fed North America and Europe, they are an asset to the active people of Amazonia, who eat few foods processed with sugar and fat. Another advantage of tucumã is that some trees are always ready for harvest, especially between April and December.

Tucumã has been planted on some of the more elevated portions of the Amazon floodplain for a long time. On Combu Island near Belém, for example, the parents of a child snacking on a fruit of *Astrocaryum vulgare* explained that the palm had been planted by the little girl's grandparents on the highest portion of the floodplain: the bank of the Combu River that meanders across the island. I found another planted tucumã in a home garden on a high bank of the floodplain at Urucurituba, near Alenquer, Pará. At the time of my visit in 1993, the garden had not been flooded since 1971. Similarly, tucumã palms in the vicinity of Santana do Ituqui, on the edge of the floodplain some fifty kilometers downstream from Santarém, appear to have been planted, or to have sprouted from discarded seed. In some cases, the palms may be descendants of palms planted hundreds or even thousands of years ago.

Yellow Mombim

Floodplain forests also harbor several other fruit trees much appreciated by locals as well as city folk. Yellow mombim, for example, produces abundant orange-yellow fruits with an exquisite tart flavor (figure 3.8). Known as *taperebá* to native Amazonians, or *cajá* to people from other regions of Brazil, the damson-sized fruits are typically mixed with water and sugar to make a delightful, bracing drink. Many a fisherman has paused under a yellow mombim tree to gather the fruits and mash them with his hands in a bowl fashioned from a calabash gourd. River or lake water is added for an instant drink.

During the fruiting season, which stretches from December to April, the fruits are marketed in riverside towns, such as Macapá. In urban areas, the fruit sells briskly to make refreshing ice creams and sherbets as well as juice. According to a rural inhabitant of the lower Tocantins, paca and tapir, both prize game animals, compete with humans for the succulent fruits. Bats are apparently also involved in dispersing the fruits, and spider monkeys drop some of the seeds away from parent trees.

Yellow mombim is widespread in the humid tropics of Latin America, where it has become naturalized in many places in both uplands and along floodplains. As people abandon home sites, yellow mombim becomes feral. It is therefore difficult to tell which populations are truly wild, and which are escapees. The plant's origins are obscure because people have moved it around for millennia. Wild populations of yellow mombim are more common on the floodplain of the Amazon and some other rivers, such as the Xingu, and is likely indigenous to the myriad

Figure 3.8. Fruits of yellow mombim gathered on the floodplain. *Taperebá*, as the fruit is known in the Brazilian Amazon, was the most abundant fruit in the market of Macapá, Amapá, in December 1994. The tart fruits are used to make refreshing drinks and ice cream.

Figure 3.9. Gathering bacuri (*Rheedia brasiliensis*) on the floodplain near Santarém, Pará, May 1996. The floodplain bacuri does not appear in markets and is therefore one of the many "phantom" fruits that do not turn up in official statistics. It is nevertheless much appreciated locally and is a regional fruit with economic potential. Other fruit-bearing species called bacuri are common in certain second-growth areas in upland areas of Pará.

waterways of the region. Yellow mombim is sometimes found in groves, such as near Itacoatiara, probably the result of deliberate planting or spontaneous germination from seeds discarded by people. According to Protásio Frikel, an eminent anthropologist formerly of the Goeldi Museum in Belém, the Tiriyó attribute unusual concentrations of yellow mombim in upland forests in the watershed of the Paru, a tributary of the Amazon, to plantings by their ancestors.[4]

Yellow mombim probably spread from floodplains to uplands, either by planting, or because it colonized light gaps created by farming activities. In upland areas it occurs spontaneously on better soils, such as the alfisols (*terra roxa*) in the vicinity of Medicilândia along the Transamazon Highway in Pará. In that regard it is analogous to the kapok tree, which is normally confined to floodplains but occurs sporadically on upland sites where the soils are unusually fertile.

Bacuri

Several species of bacuri are cultivated or gathered in the wild in the Brazilian Amazon, but only one is at home on the floodplain: *Rheedia brasiliensis*. This species, which grows along forest margins where it receives a lot of sun, sports bright-yellow fruits in profusion from spreading branches during the annual flood (figure 3.9). Bacuri trees are nearly covered at the height of the flood and fish can literally pluck the Ping-Pong-ball–sized fruits from the branches. The floodplain bacuri is consumed locally and rarely appears in markets. The fruits are sweet-tasting—not too acid or tart—so they are appreciated as a snack food as well as for juice. One afternoon in May 1996, a boat I was traveling on from Arapixuna to Santarém stopped at a partially submerged bacuri tree. After the crew gathered a bucket of the soft-skinned fruits, the captain resolved it was time to leave. "We'll leave the rest of the fruits for the fish," he declared. While I was not entirely convinced of his conservationist ethic, his remark was at least evidence of the intimate knowledge of natural history of the people who live and work on the floodplain.

The Tree Legumes: Marimari and Ingá Baú

To temperate-zone dwellers, the notion of legumes as fruit seems odd; the family is usually associated with vegetables and pulses. But in the tropics, many species of the legume family are forest giants. Leguminous trees are common on the Amazon floodplain, and two denizens of *várzea* forest yield much appreciated fruits. Along the floodplain of the middle Amazon, marimari and ingá baú trees are harvested in the wild for their pendulous fruits, which mature during the flood stage.

The corrugated, twisting pods of marimari can reach a meter or more in length and contain a slightly sweet, white pulp around the seeds (figure 3.10). When the pods turn yellow, the fruit is ripe, but they are often picked green, before their prime. The location of marimari trees along the sunlit forest edge is well known to locals who maintain "mental maps" of the natural resources in their surroundings. The fruits are harvested by paddling a canoe right into the branches and then twisting the pods until they break free.

Figure 3.10. Pods of marimari, gathered in floodplain forest, for sale in Juriti, Pará, June 1994. The sweet, white pulp surrounding the seeds is much appreciated by people in rural and urban areas. The flag on the wheelbarrow indicates that the fruit is for sale.

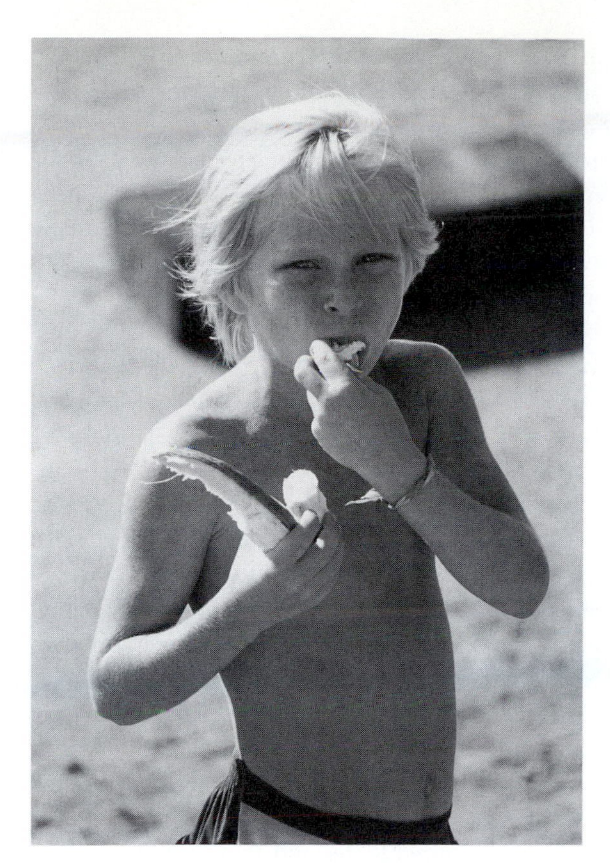

Figure 3.11. The fruits of ingá baú, which are harvested from the wild on the floodplain, are a welcome snack along the middle and lower Amazon. The boy's blond hair suggests a strong European ethnic heritage; his playmates range the entire spectrum from indigenous to African and European ethnic origins. Óbidos, Pará, June 1994.

Marimari finds its way to markets from April to June in such widely scattered towns and cities as Itacoatiara, Juriti, Manacapuru, Manaus, Óbidos, and Santarém. In Santarém in June 1994, bundles of marimari fruits were selling for about seventy cents a kilogram, a relatively high price for the region. Upstream at Juriti, the fruits were even more expensive at eighty cents a kilogram, but at Óbidos, where marimari may be more common, they were selling for fifty cents a kilogram during the same month. Only a small percentage of the fruit by weight is edible, typical of wild fruits. Some farmers have deliberately planted marimari in their backyards. The next step will be selection for trees that produce a greater proportion of pulp.

Ingá baú, also known as ingá açu (the big ingá named after its generous, almost cylindrical pods), is characteristic of river banks along white waters in the Amazon basin and occurs down the Amazon all the way to the estuary. It has shorter, thicker pods with more generous quantities of pulp surrounding the larger seed (figure 3.11). Ingá baú is sold in street markets in Santarém and Óbidos. In the latter town, the thick, woody pods of ingá baú were being sold for forty cents a kilogram in June 1994.

Sapucaia

People also avidly seek nuts in floodplain forests, especially those from the sapucaia tree. In the same family as the better-known Brazil nut, sapucaia grows on both the Amazon floodplain and in widely scattered locations in upland forest. In interfluvial forests, such as between the Tocantins and Xingu, sapucaia emerges above the canopy and people are able to savor the nuts only when a windstorm knocks down some of the pods from the forty-meter-tall trees. On the Amazon floodplain, however, sapucaia is shorter and looks more like a spreading chestnut tree; the pods can thus be readily dislodged with a knife tied to a long pole. Thick, spreading sapucaia trees provide generous shade, and are commonly planted or spared from the ax around houses.

Pollinated by carpenter bees, sapucaia flowers in the dry season and its nuts are harvested during the rainy period from January to May. Also called *castanha* in the vicinity of Santarém, sapucaia nuts are encased in a large, bowl-like shell (called a capsule by botanists) that hangs upside down (figure 3.12). The tasty nuts are secured by a fleshy appendage (an aril to botanists) and are tamped inside the inverted "bowl" by a saucer-shaped lid. When the nuts are ripe, the lid falls off and fruit-eating bats abscond with the nuts to feed on the succulent arils. After the bats have finished dining, the nuts are dropped to the ground where some eventually germinate. To subvert the bats or prevent the nuts from spilling on the ground, people frequently harvest the nuts just before they are fully ripe.

Sapucaia nuts are consumed locally and sold in nearby markets. In the latter part of the nineteenth century, small quantities of sapucaia nuts made their way to the London market as a luxury item. In the early 1900s, some 6 metric tons of sapucaia nuts were exported, mostly to the United States where they were called paradise nuts, a savvy marketing ploy. The advent of regular steamships plying the Atlantic between the Amazon the United States and Europe a little over a

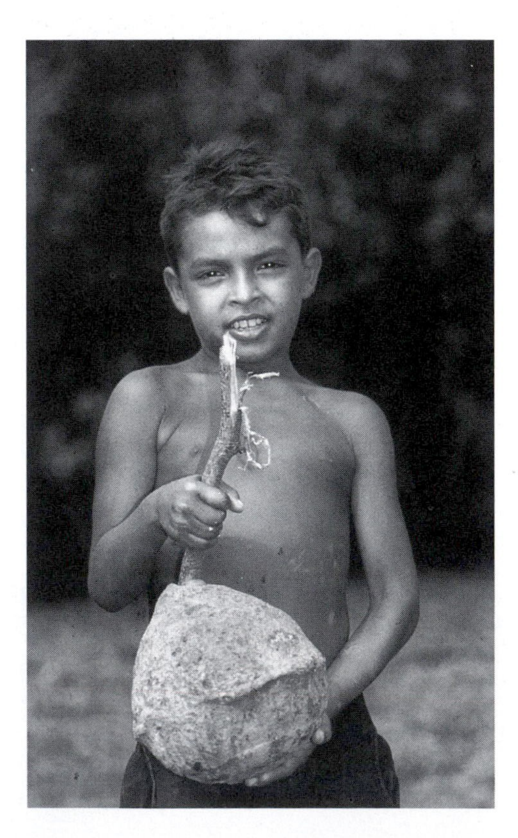

Figure 3.12. A sapucaia capsule full of nuts from the floodplain near Santarém, Pará, May 1996. The boy is holding up the capsule by its stalk, and the nuts will be accessed by chopping open the lid near his lower hand.

Figure 3.13. Sapucaia capsules used for flower pots. The mother is of indigenous ancestry and an avid gardener. Juriti, Pará, June 1994.

century ago made it possible for perishable sapucaia nuts to be exported. Curiously, these nuts are no longer exported.

Sapucaia capsules are sometimes used locally as flowerpots (figure 3.13), and are also sold in urban centers for ashtrays. Some even make their way to gift shops in England; in Leyburn, Yorkshire, they were retailing for five dollars each in 1993. In the mid-nineteenth century, the capsules were sold in London's Covent Garden as "monkey's drinking-cup."

Camapu

Disturbed sites around fields and homes, as well as the ridges and swales of freshly formed land, represent another hamper of foodstuffs. Around fields and along the margins of lakes in the vicinity of Itacoatiara and Santarém, for example, people snack on the husked fruits of camapu, a weedy herb (figure 3.14). Camapu is Amazonia's answer to tomatillo; for aficionados of Mexican cuisine, tomatillo is a major ingredient in *huevos rancheros*. In Amazonia, camapu fruits sometimes turn up in urban markets, such as those in Macapá, where it is used in a dessert called *doce de camapu*. A relative of the tomato, waist-high camapu is a weed or a resource, depending on one's perspective. To modern, commercial vegetable growers on the Amazon floodplain, the bushy herb can be a nuisance; to others, the husked fruits of the volunteer plant are a welcome supplement to the diet.

It is worth remembering that today's weed can be tomorrow's crop; plants can slip in and out of domestication as long as they are not totally dependent on humans for reproduction. Two near relatives of camapu have been domesticated: one in Mexico (*Physalis philadelphica*), called tomatillo or tomate verde; the other (*P. peruviana*), in the Andes. The Andean species is no longer grown, but it still occurs wild and the fruits are gathered. In Mexico, tomatillo was domesticated in

Figure 3.14. The husked fruits of weedy camapu are a snack food on the floodplain. The young man, an agricultural laborer paid by the day, has just picked a fruit of camapu growing near a vegetable plot; other fruits can be seen on the herbaceous shrub to his left. The worker sports a torn T-shirt advertising a paint store in Santarém. Piracauera de Cima near Santarém, Pará, November 1992.

Pre-Columbian times to complement sauces and is both cultivated and gathered in the wild. One day, with changing tastes and market conditions, vitamin C–rich camapu could also become a crop along the Amazon, even switching places with the tomato.

Camapu may never follow the same path as its close relative in Mexico, but its use as a snack food along the Amazon serves to illustrate the value of closely scrutinizing all plants used by people in their environment. Most of the world's major crops originated in open environments. More crop candidates surely await discovery on the Amazon floodplain, an area with abundant, ephemeral light gaps on varied terrain, ranging from lake margins to newly formed beaches.

Wild Rice

Floating meadows cover substantial portions of lakes, placid side-channels, and small bays along the banks of the Amazon, especially at high water. They contain several food plants, now mostly underutilized. In the past, riverine folk and hungry rubber tappers harvested wild rice from buoyant grasslands. The golden grains were simply shaken directly into canoes or baskets at high water. Today, few people harvest this close relative of cultivated rice. Its only current use along the middle stretches of the Amazon is to feed captive capybara, the world's largest rodent. A farmer near Santarém who is raising these animals for food gathers wild rice and other floodplain grasses to feed them.

Floodplain Forests as Bait Shops

Virtually all farm families fish at some time of the year, and fish sales are a major source of income for many of them. Given the importance of fish for subsistence and commerce, it is not surprising that a large assortment of wild plants are tapped to use as bait. Numerous floodplain trees and bushes attract fish. "Fish-bait" trees generally fruit at high water, since many of them depend on fish for seed dispersal. Along the middle Amazon from Itacoatiara to Santarém, fisherfolk comb forests for more than a dozen fruits to bait their hooks.

Palms are conspicuous in landscapes along the Amazon and many of them—such as the jauari, the marajá, the pupunharana, and the urucuri—anchor fish food chains. Many other trees along the floodplain produce fruits that are gathered for fishing: for example, catauarí, uruá, rubber tree, socoró, tarumã, cecropia, caimbé, cramuri, and arum.

Jauari Palm

People collect the fruit of jauari palm for fish bait as well as to feed pigs. Along the Paraná Nhamundá, one fisherman reported that he uses the entire golf-ball–sized fruit to catch tambaqui, a fish that can weigh as much as thirty kilograms and easily fetches the equivalent of a month's minimum wage at market. He says that he catches the smaller aracu by baiting the hook with pulp scraped from the hard, wood-encased nut. Striped aracu do not command the same high price in fish markets as tambaqui, but poorer urban residents find them more affordable. Some stands of jauari have fallen victim to heart-of-palm factories in Manaus, and others have been cleared for cattle ranches and crop farming. The once dense stand of jauari palms on the eastern outskirts of Itacoatiara, Amazonas, has succumbed to urban sprawl. The Jauari district of Itacoatiara, site of one of the main fish markets for the growing town founded over two centuries ago, did not contain a single jauari palm by the late 1970s.

Marajá and Pupunharana Palms

The relatively short marajá palm is also a favorite of many who fish, at least from Careiro Island near Manaus to Paraná Nhamundá near Oriximiná at the confluence of the Amazon and Trombetas. Fishermen lop off the fruit stalks of the widespread palm, gaining a dozen or more purple fruits, and place the fruits in the bottoms of their canoes (figure 3.15). Once at the fishing grounds, they place the marble-sized fruits on hooks to capture a wide assortment of fish, including pirapitinga and tambaqui. The fruits of pupunharana, which resembles the domesticated peach palm, are also collected to fish for tambaqui.

Urucuri Palm

In the vicinity of Juriti on the south bank of the Amazon, urucuri palm is used in a most unusual manner to catch fish. The fruit of urucuri (figure 3.16), which was once used to smoke the latex of rubber trees, is first scraped to remove the pulp. The pulp is then tied into a ball with twine and secured to a line. Four to

Figure 3.15. The fruits of marajá palm are used as fish bait along the middle Amazon. On the right side of the canoe are several fishing poles and a pacu fish can be seen by the machete. Paraná Nhamundá near Terra Santa, Pará, June 1994.

Figure 3.16. The urucuri fruits are to be used for catching jaraqui, which pass up the Amazon in large schools as the flood subsides. The fruits of the palm are also gathered to feed pigs. Juriti, Pará, June 1994.

five unbaited hooks are attached below the ball. Above the bait, the line extends for a meter or two to a float, usually a piece of Styrofoam. With the line secured in his hand, the fisherman sits in his canoe on the river. When trout-sized jaraqui feed on the pulp, the float bobs, which is the fisherman's signal to yank the line hard and sometimes snag a jaraqui or two. The "snagging" method to catch jaraqui is also reputedly used along the Arapiuns River and in the Santarém area. Because the small-mouthed jaraqui are "rasping" feeders and will not take to bait on a hook, they are normally caught in seines or castnets during their annual migration from the Amazon floodplain as the water level drops.

Catauarí

The green-gray fruits of catauarí are another favorite food of tambaqui, so farmers and fishermen from Itacoatiara to Monte Alegre, and probably beyond, remember from year to year the location of catauarí trees. Other fruit-eating fish caught with the orange-sized fruits of catauarí include round, silvery pirapitinga, which can weigh up to twenty kilograms. Catauarí has one of the longer fruiting seasons of the fish-bait trees: from November, at the tail end of the low-water period, to July when the flood begins to recede.

Uruá

Uruá is another floodplain tree with fruit that is avidly sought by tambaqui, pirapitinga, and aracu. Tan-colored fruits of uruá occur in bunches and can be easily slipped on a hook (figure 3.17). The blueberry-sized fruits are a valued fish bait along the Óbidos-Santarém stretch of the Amazon.

Rubber

Tambaqui are also partial to rubber seeds, which are gathered from the ground in large quantities for sale to urban-based fishermen. In the past, the shiny seeds served as a snack food for people, but apparently this practice is now rare. Rubber was formerly planted extensively on the Amazon floodplain and on nearby upland bluffs for latex production, but now the trees are rarely tapped for their milky latex. Most of the rubber seeds used in fishing come from trees on the uplands, because floodplain trees drop their seed at high water and fish eat most of the seeds. Rubber trees produce an explosive popping sound as the dry capsule bursts open and releases the seeds, thereby attracting frugivorous fish.

Socoró and Tarumã

Riverine inhabitants relish the red, cherry-like fruits of socoró, another floodplain tree used primarily for fish bait and much favored by tambaqui. Socoró has a relatively brief season, fruiting most heavily in April and May as the waters are still rising. By the time the flood crests, most of the trees have finished bearing their fruit. Tarumã fruits are collected as bait for pirapitinga and jatuarana and occasionally serve as snacks while paddling in floodplain forests.

Figure 3.17. Fruits of uruá are used as fish bait, particularly to catch tambaqui. Uruá trees are typically spared when farmers clear home sites on the floodplain, and spontaneous seedlings are protected in home gardens. Amador near Óbidos, Pará, June 1994.

Cecropia

Although most of the fruits used for fish bait are picked in the floodplain forest and along its margins, some plants that colonize disturbed areas, or new islands and bars, also produce fruits that attract desirable fish. The sun-loving cecropia tree, in particular, produces catkin-like fruits that are snapped up by sardinha and other commercially valuable fish when they fall into the water. Known as *embaúba* in Brazil, cecropia often forms pure stands in light gaps; fishermen near Itacoatiara sometimes tarry in such spots to gather the fruits and capture their unsuspecting quarry.

Caimbé and Cramuri

Turtles are a delicacy along the Amazon and they fetch handsome prices in urban markets, even though it is illegal to sell wildlife products. Fisherfolk have thus devised a number of strategies to capture their elusive prey, and some of them involve forest fruits. In the vicinity of Itacoatiara, for example, fishermen bait trotlines—four to eight hooks suspended from a line so that they rest just below the water's surface—with the blue-black fruits of the caimbé tree. The trotline is set up at high water under branches of the small tree to catch the yellow-spotted Amazon turtle (*tracajá* in the Brazilian Amazon). Caimbé fruits until the end of

July when the floodwaters have normally been receding for about a month. The fruits of the cramuri tree are also employed as trotline bait for the yellow-spotted Amazon turtle but are available only until May, when the waters are still rising.

Arum

The fruits of arum, a giant aquatic plant belonging to the same family as Philodendron, are also deployed to attract the yellow-spotted Amazon turtle. Arum sports distinctive heart-shaped leaves and is sometimes found in dense stands along the margins of channels and in lakes of the Amazon. Its crumbly yellow fruits are used as "chum" to entice the agile turtles within striking range of harpoons (figure 3.18). When the wary turtle, which can yield several kilograms of meat, approaches the canoe, the fisherman jabs its hard shell (carapace) with a special harpoon called a *tapuá*, and hauls in his quarry. This technique is more effective in less turbid waters, such as backswamps and flooded forest, than in the creamed-coffee–colored Amazon.

In the past, some indigenous groups would wait in arum stands to harpoon the giant river turtle, which can attain fifty kilograms. Known as *tartaruga* in the Brazilian Amazon, this once abundant species has become so rare that waiting in arum groves is no longer a worthwhile proposition. Overhunting, exploitation of its eggs, and elimination of many of its food resources with deforestation on the floodplain, have taken their toll on these "river cattle" that were once corralled and fed arum leaves until ready for the table.

Figure 3.18. The pineapple-like arum fruits in the canoe, known locally as aninga, are broken into pieces and tossed into the water to attract turtles within striking range. The man, a descendant of African slaves, is waiting for turtles to appear. Igarapé do Lago, an affluent of the Maracá, Amapá, May 1996.

A Living Hardware Store

Fisherfolk also employ a variety of plants from the Amazon floodplain to fashion gear. The long, graceful canes of flecheira, a bamboo-like grass, are cut along the banks of the middle and upper Amazon to make arrow shafts that float. Bamboo splits easily, so fishermen attach a hard "stopper," fashioned from tucumã palm, at the end of arrow shafts upon which the metal point rests. Over a century ago, an intrepid English geographer, William Chandless, noted that sturdy flecheira was used to pole canoes along the meandering Purus, a white-water affluent of the Amazon.[5] Jacques Huber, a field botanist working out of the Goeldi Museum at the turn of the twentieth century, asserted that flecheira first appears around Parintins as one travels up the Amazon.[6] Evidently the range of the cane has since changed, since I saw it in flower in July 1996 on Capivara Island, about midway between Santarém and Óbidos. Flecheira is especially common along the floodplains of the Marañon and the Ucayali in the headwaters of the Amazon.

The pliable inner bark of munguba and sapucaia are used to string fish and to tie the restless feet of yellow-spotted Amazon turtles so that they do not escape or rake their captors with their sharp claws. Fishing poles are fashioned from the branches of abiuarana (unidentified), the stems of marajá palm, and several trees from upland forests.

The floodplain of the lower Amazon is home to jupati, a palm with the world's largest fronds. Jupati's fronds can arc for ten meters; when one enters a sun-filtered grove of the jupati palm images of prehistoric landscapes come to mind. The mid-ribs (*talha*) of the generous fronds are used to fashion shrimp traps (*matapís*), sold in various fishing villages and towns throughout the Amazon estuary (figure 3.19). Shrimp traps are also made from the long stems of guarumã (called uarumã or urumã in northern Marajó), an understory plant from a different family: the Marantaceae.

Interestingly, locals turn to an upland palm to bait shrimp traps. Floodplain residents purchase nuts of babaçu, a relative of the urucuri palm, to entice freshwater shrimp into the traps. Almond-sized babaçu nuts are pounded into small pieces and wrapped in leaves or plastic; in this manner the scent disperses in the water but the nuts are not easily eaten or washed away.

In the past, several floodplain forest trees, such as favarana and jacaréuba, were used extensively along the middle Amazon to fashion canoes and build boats. Now with so many of the floodplain forests felled and burned or logged off, boat builders have been forced to buy durable wood from suppliers along the Upper Amazon, or on the uplands. In the middle Amazon at least, most craft are now built with the deep, red bow-wood or tan-colored itaúba, both giants of upland forests. In the estuarine area, such as on Combu Island, paddles are carved from the buttress roots of the matutí tree.

As in North America and Europe, the well-to-do in the larger towns in the Amazon buy their construction supplies, kitchen utensils, and garden tools from a hardware store or supermarket. But the income of most residents of the Amazon floodplain is limited, so they must garner as many supplies as possible from their immediate environment. Men, women, and children paddle or walk into forests to gather wood, twine, and fronds for a wide variety of purposes, including the construction of houses and nurseries for seedlings.

Figure 3.19. Shrimp traps fashioned with fronds of jupati palm, gathered in estuarine forest of the Amazon. The traps, which are typically baited with crushed nuts of the babaçu palm, rest on sleeve-like presses called tipitis, used for squeezing manioc dough. More such presses can be seen stacked in the background. Mats made from junco, a floodplain rush, are next to the salesman. Mosqueiro, Pará, June 1994.

Even the simplest houses require some wood, if only for corner posts. A favored tree for the main support posts is mulato wood, now increasingly rare on the floodplain, to which it is confined. Mulato wood is also favored for plank floors. Until the mid-nineteenth century, mulato wood was abundant along the Amazon up to the footslopes of the Andes, but it was then heavily cut until about 1930 to supply fuel to steamships. Loggers can easily spot the tree with its distinctive peeling outer bark that reveals a smooth, red trunk.

The piranha-tree, with its extremely hard wood that resists termites and prolonged waterlogging, is preferred for stilts to raise house floors above floods. But it, too, has become difficult to find because of rampant deforestation and uncontrolled logging. On Ilha Grande near Óbidos, a house that is occupied year-round has a raised floor of andiroba planks. Andiroba is a sizable tree of floodplain forests with several medicinal uses. But given the scarcity of top-quality timber trees on the middle Amazon floodplain, many of the floors and ceilings now consist of planks sawn from trees that would otherwise be saved for medicinal or other purposes.

In the Amazon estuary, fronds of the bussú palm have long been the material of choice for roofs for those who cannot afford tile or corrugated asbestos sheets. The fronds of bussú reportedly last for ten to twelve years after cutting, compared to two or three years for most of the other palms in the region. With its generous fronds that spray in all directions, bussú is a distinctive feature of floodplain forests at the mouth of the Amazon, where it is also known as *palheira*. Small boats laden almost to overflowing with the gray-green fronds of the palm are a familiar sight plying the twisting, mangrove-lined channels of the estuary. Bussú is so highly regarded that rural people living along the tributaries of the lower Amazon, such as the Anapú, sometimes send in "orders" to suppliers for the fronds. It is also in demand for roofing in the poorer districts of urban centers; in August 1988, I witnessed a boat off-loading bussú at Tucurui on the Tocantins. In November 1995, bussú fronds were selling for one hundred dollars per thousand at dockside in Belém. Small quantities of the dark brown, fibrous veil that partially covers the flower stem of bussú is also gathered. Called *tururi*, the sack-like material is fashioned into handbags and collapsible hats for the tourist trade.

Along the middle Amazon, where bussú drops out of the picture, floodplain dwellers generally resort to upland palms to thatch their houses. In the vicinity of Santarém, curuá is the cover of choice for houses. Floodplain residents do not have to go deep into the jungle to secure curuá fronds; the palm is abundant in disturbed sites on upland bluffs overlooking the Amazon. In fact, curuá is so abundant around Santarém that a river that drains into the Amazon from the south is called the Curuá-Una (black curuá), now dammed at the first rapids. And on the opposite side of the Amazon between Alenquer and Óbidos, the still-free Curuá River enters the Amazon by the village of Curuá.

Several items used in and around the house are also derived from trees growing in floodplain forests. Brooms are fashioned from the inflorescences of açaí palm. At the mouth of the Amazon, baskets are fashioned from the fronds of buriti and jacitara palms to carry açaí fruits. Jacitara is found in higher parts of the floodplain, such as on Marajó Island, as well as in upland forests. Baskets are sometimes lined with the large, fleshy leaves of sororoca, a banana-like plant of the understory in floodplain forests, especially if the weave is relatively open. Sororoca occurs in small groves in the dark interior of moist forests; the untrained eye might suspect that one is in the midst of an overgrown banana plantation. But sororoca does not produce fruits like banana; only its stems and leaves are collected.

Long-stemmed guarumã is also employed to make baskets for açaí fruits, called *rasas*. Guarumã is an understory shrub of floodplain forests of the Amazon and some of its tributaries, such as the Maracá, and also grows along streams in some upland areas of Pará. Guarumã is also woven into mats to dry cacao beans (figure 3.20). Near Santarém, rural folk employ it to fashion *tipitís*, sleeve-like sieves for squeezing manioc dough. One farmer who lives near Alter do Chão, some thirty kilometers southwest from Santarém, complained that deforestation by ranchers and migrant settlers in the area has made guarumã scarce. Tipitís are also made from the fronds of buriti, such as in the vicinity of Juriti on the Amazon and Caxiuanã on the lower Anapú.

Buriti also figures in religious ceremonies. At Terra Santa along Paraná Nhamundá, for example, women carve the palm's soft wood to make model boats in preparation for the Círio Santa Isabel, a local festival sponsored by the Catholic Church. Each year at the end of June, the faithful launch twenty-centimeter-long boats, laden with glowing candles and offerings to the Virgin Mary, onto the floodplain lake in front of Terra Santa.

Cattle ranching is the major land use along the middle Amazon, and several wild plants are enlisted to facilitate life on the ranch. A rush, known locally as junco, is collected from marshes to make a thick pad that is placed underneath saddles (figure 3.21). Curtains of freshly cut junco hung out to dry are a familiar sight on ranches from Nhamundá to Santarém. Thick pads of junco are sold for

Figure 3.20. A guarumã mat used for drying cacao beans. Guaramã, also used to make baskets, is an understory shrub in the floodplain forest. Combu Island near Belém, Pará, December 1994.

Figure 3.21. Saddle pad made from junco, a reed gathered on the Amazon floodplain. Junco is also fashioned into mats to protect produce at open air markets in towns along the middle Amazon. Fazenda São Lourenço, Paraná Miri near Alenquer, Pará, March 1993.

saddle padding in stores and markets along the middle Amazon, such as Alenquer. Ranch hands often fashion their own saddle pads from junco, as on Marajó Island. In Santarém, large mats of junco are used to cover produce stalls in open-air markets to help protect fruit and vegetables from the midday sun.

Small-scale ranchers are often farmers as well, and they need fences to keep cattle away from their crops, or to maintain good neighborly relations. At Murumuru near Santarém, one farmer employs a living fence of yellow mombim to keep his handful of cattle out of neighbors' crops, a practice also noted in Costa Rica. Yellow mombim thus provides a double blessing: crop protection and succulent fruits for juice making.

The emergence of vegetable farming as a significant economic activity along some parts of the Amazon, particularly near urban centers, has created a demand for "garden supplies," including fencing material. Some vegetable growers opt for living fences, using yellow mombim or the sun-loving Amazon willow. The Amazon willow, known in the Brazilian Amazon as *oeirana*, colonizes new mud flats along white-water rivers, often forming almost pure stands. It is so common that in some areas channels and streams are named after the quick-growing tree, such as Paraná Oeirana on Carmo Island between Santarém and Óbidos. Farmers, such as those on Tapará Island near Santarém, cut branches of the lime-green foliage and stick them in the fertile alluvial soil, where they readily root and soon form a tall fence. The Amazon willow is also used to stake tomatoes, as along Paraná Ituqui and Urucurituba in the vicinity of Santarém (figure 3.22).

Figure 3.22. Tomato stakes fashioned from Amazon willow. The girls help their parents with farm chores after school. At high water, most of the farmland is under water and schools operate precariously. Boca Aritapera de Cima, near Santarém, Pará, November 1992.

Figure 3.23. A firewood stove in a kitchen of a floodplain home. The Amazon has recently spilled into the backyard, and will keep on rising for a couple of months. Although the house is on stilts, floodwaters pour through the house in some years. The teenager is wearing a T-shirt promoting a candidate for vice-mayor in Santarém; the slogan reads "you're the winner." Igarapé-Açu near Santarém, Pará, March 1993.

To keep costs down, some farmers turn to mulch derived from the trunks and branches of dead munguba trees to prepare beds for seedlings. Munguba, which is confined to white-water floodplains, sports distinctive red pods full of wind-dispersed seed at high water. Cork-like munguba wood is also used in vegetable gardens, both raised and on the ground. Rotten munguba provides easily worked material that encourages root development and provides organic matter to the soil; its spongy texture also retains soil moisture and helps suppress weeds. I have seen munguba thus employed in various locations on the Amazon floodplain or adjacent to it, including Tapará Island and at Carariacá, both near Santarém.

Vegetable growers also tap another common floodplain tree, the urucuri palm. Partial shade for tender vegetable seedlings is provided by cutting fronds of urucuri palm and laying them on supports over the nursery beds. Urucuri fronds also protect delicate seedlings from pounding rain.

Fuelwood for the Kitchen and Manioc House

Most people in industrial countries care little about the source of energy used in cooking. Electric or gas ranges are taken for granted; heat is virtually instantaneous. The middle and upper classes in towns and cities in the Amazon rely almost exclusively on bottled gas to cook their meals. But an urban bias can blind us to the fact that most of humanity still depends on fuelwood or charcoal for cooking.

In rural areas of the Amazon, firewood and charcoal are the primary cooking fuels (figure 3.23). The search for enough wood to prepare meals is thus a daily chore for most residents of the Amazon floodplain and uplands. One can conceive of fuelwood shortages in dry regions, but it may be difficult to imagine a fuelwood crisis looming in the well-watered Amazon basin. Yet as forests along the Amazon continue to shrink, fuelwood supplies are likely to emerge as a resource issue.

Fuelwood is gathered from a variety of sources including floodplain forests, partially burned trees in swidden plots, driftwood left by receding floods, and to a lesser extent from trees planted or left in home gardens. Not all woods have the same heating value and certain species are preferred for cooking and to prepare manioc flour. Sought-after fuelwood on the floodplain in the vicinity of Alenquer and Santarém includes mulato wood, paricá da várzea, sapupira, and various species of ingá.

Manioc flour is a staple throughout the region and requires appreciable quantities of fuelwood for toasting it. It takes about 3 hours to transform manioc dough to gritty flour on a metal griddle. Slow-burning wood with a high heat value is thus preferred for the oven underneath the griddle. On Careiro Island near Manaus, such woods include acapurana, espinheiro, limorana, piranha-tree, and tarumã. Further downstream in the vicinity of Oriximiná and Santarém, mulato wood and tarumã are considered among the best fuels for a manioc oven. Many residents make manioc flour at low water to sell in nearby towns and cities.

Conservation of crop genetic resources and fisheries should be central to any overall plan to develop and manage natural resources on the Amazon floodplain. But little attention has been paid to how future generations will cook their meals or prepare manioc flour. Shortages of fuelwood have already emerged as a constraint to farmers who wish to process their manioc into flour in parts of the Bragantina zone east of Belém. The uplands east of Belém have been colonized for more than a century and most of the forest is now gone. The day may not be far off when fuelwood shortages crop up along stretches of the Amazon. The importance of fuelwood to the well-being of floodplain residents underscores again the imperative task of protecting and managing wisely forests along the Amazon. In so doing, a plentiful stock of flood-adapted candidates will be available for planting on a larger scale to supply fuel.

A Forest Pharmacy

Most people who live on the Amazon floodplain can ill afford prescription medicines, let alone the fees of private doctors. Public health care is available generally only in the larger towns and cites and entails a time-consuming boat trip. Given the huge demand for understaffed public health services, long lines are common, and patients may have to stay overnight in order to secure an appointment. In view of the expense and inconvenience of conventional medical care, many floodplain residents resort to numerous wild and domesticated plants when they feel sick.

Knowledge about the real or imagined therapeutic properties of plants is passed down through the generations, particularly from mothers to their daughters. If a health problem is particularly perplexing, people seek the advice of a local healer (*benzedor, curandeiro, pajé*). Curers are often women who have acquired much specialized knowledge about the pharmacological value of certain plants. They rarely charge a fee for their instructions or preparations, relying instead on donations, goods, or services. Some of the more famous healers set up practice in towns or cities, but are still cheaper than conventional medicine.

Curers do not see themselves in opposition to modern medicine. In cases of severe injury or serious illness, such as polio, patients are advised to go to the nearest hospital or clinic. And most floodplain residents maintain a stock of store-bought remedies, such as antibiotics or syrups for strengthening the liver, typically bought without a prescription. In many cases, such medicines are kept beyond their expiration dates. The pharmacopoeia of rural inhabitants encompasses both laboratory-derived medicines and folk remedies based mostly on plants.

Some of the plants used in folk remedies are cultivated around homes, but others are gathered from the wild. Most of the wild plants used in domestic treatments are gathered from floodplain forests, where the greatest diversity of plants is found. Numerous forest plants are used to treat a wide assortment of health problems ranging from diarrhea to intestinal worms and fever. I only note a few here. Several floodplain trees, including assacu-rana and louro-inhamuí, are employed in remedies for liver disorders. Several trees are helpful in cleaning and disinfecting wounds and skin disorders, particularly andiroba, caimbé-rana, the floodplain tortoise tree, and paracaxi. Albina is a useful expectorant, while the latex of caxinguba purges hookworm. Both cumacaí and jipioca are reputedly effective in treating dandruff.

Medicinal plants often have multiple uses. A salve for skin lesions, for example, is prepared from the sap of virola, a valuable timber tree. Oil from the fruits of the American oil palm is used to soothe whooping cough. Several wild fruit trees also double as medicinal plants. The inner bark of yellow mombim is used to make a tea for treating diarrhea on islands between Santarém and Alenquer. The roots of the açaí palm are boiled and the resulting tea is taken to treat swelling and intestinal worms. Andirá-uxi is also a vermifuge according to informants on Ilha do Carmo near Santarém. Leaves of assacu, the preferred tree to provide logs for floating cattle corrals, are used to make a decoction to treat various inflammatory disorders. According to the ethnobotanist Richard Evans Schultes, Peruvians prepare a decoction with munguba, a floodplain tree widely used by fisherfolk and farmers alike, to treat snakebites.[7]

It is sometimes argued that folk remedies eventually will be surpassed by the spread of modern medicine, and that traditional medicinal practices—and the plants upon which they are based—are a thing of the past. But the pharmaceutical value of floodplain forests, and other tropical rainforests, could eventually prove one of their major assets. It is safe to say that less than 1 percent of Amazonia's plant species have been screened for their therapeutic value. Some major pharmaceutical companies have stepped up their prospecting efforts in tropical forests, but much remains to be done. The search for new drugs to treat cancers and other illnesses has attracted some small, start-up companies that are actively screening plants gathered in tropical forests for pharmacologically active compounds.

Such screening relies heavily on local knowledge of medicinal plants. The oily nuts of andiroba (figure 3.24), a towering tree of floodplain forests, have long been used by locals as an insect repellent and anti-inflammatory agent, and are now being investigated by pharmaceutical companies for their pain-killing properties as well. Other plants may also contain compounds useful for treating ill-

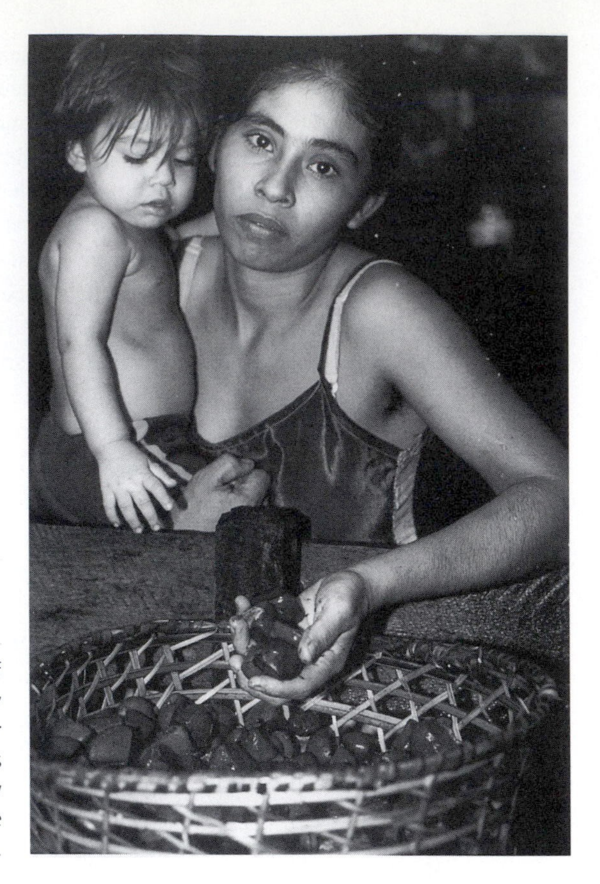

Figure 3.24. Oil is extracted from the nuts of andiroba, a floodplain forest giant, to treat a variety of health complaints. Andiroba, among other essential oils from Amazonia, is finding new markets in aromatherapy, which is gaining popularity in spas in various countries, including the United States. Rio Maracá, Amapá, May 1996.

ness, thus underscoring the importance of conserving as much biodiversity as feasible. For example, cataurí, the fruits of which are employed as fish bait along the Amazon, also acts as an antidote to snake venom.

The conservation of tropical forests and other plant communities along the Amazon River is critical to humanity's future ability to combat disease on two counts. First, drug companies need a large stock of plants to test for pharmaceutical properties. Second, even after a suitable compound is identified and synthesized in the laboratory, intact forest and other natural vegetation is essential for continued evolution. The numerous agents that trigger disease in humans, such as bacteria, fungi, and viruses, mutate frequently. New medicines must then be devised to combat them.

The Timber Industry

Whereas in the past, logging was concentrated along rivers, today most of the action in the timber industry is on the uplands. A large and growing network of pioneering highways and associated side roads have opened up vast areas of hinterland to loggers. Brazil has emerged as Latin America's largest producer of industrial sawlogs and pulp, and the Amazon has recently become Brazil's most important source of industrial roundwood. Several timber concerns in Southeast Asia are about to set up operations in the Brazilian and Peruvian Amazon, and the tempo of logging will likely quicken.

Much of the valuable timber has been logged off the middle stretch of the Amazon floodplain, roughly from the confluence of the Negro to the mouth of the Xingu. In the 1960s and early 1970s, large V-shaped rafts of logs tied together were a common sight on the Amazon near Manaus and Santarém. Today, logging on the Amazon floodplain is concentrated upstream from Tefé and in the estuarine area (figure 3.25). Upland forests within easy access of rivers have also been largely creamed of their economically valuable timber.

Even in areas where floodplain logging is still under way, signs are appearing that exploitation rates may be excessive. Factories that make veneers have favored virola, the principal timber species of the floodplain, because its large logs facilitate the shaving process and the wood does not easily split or crack. Many of the larger specimens in the vicinity of Breves on Marajó have already been cut, and trees under thirty centimeters in diameter are now typically logged. Regeneration of ucuúba, as virola is called in Brazil, is apparently suffering because of habitat disturbance associated with logging in swamp forests of western Marajó Island.

Much of the floodplain forest has been cut in the vicinity of Santarém, and most of the sawmills in town now obtain their wood from uplands. Even where patches of forest remain, most of the valuable timber has been removed. In addition, some of the softwoods, such as the kapok tree, have been culled for ply-

Figure 3.25. One of the many sawmills in the Amazon estuary. This is one of the smaller mills, catering to local demand for timber. Others are industrial-scale and are geared to extra-regional markets. Furo de Breves, Pará, November 1995.

wood factories, where they serve as "filler." Brazil has emerged as the world's third largest exporter of plywood, and countless kapok trees have succumbed to this burgeoning trade. In the past, the trees were occasionally felled to help float some of the more valuable hardwoods, but they were still relatively common until the 1970s. Along the Amazon near its confluence with the Negro, the Mura used mature kapok trees as lookout posts in the eighteenth century, and in the early 1800s, the German naturalists Johann Spix and Carl Martius noted that the tree was a frequent canopy emergent along the Amazon.[8] In the mid-1920s, an English botanist, Ruggles Gates, noted while traveling on the Amazon that kapok trees were "scattered at intervals all along the river's border."[9] The grandeur of the pillar-like kapok trees with their distinctive umbrella-like crowns and wall-like buttress roots, are now a distant memory for most residents of the Amazon floodplain (figure 3.26).

Concern is mounting that the pace of logging in upland forests is unsustainable. The rapid demise of commercial logging along the middle Amazon is a harbinger of what is to come in many upland areas. The bottom line here is that it "pays" in the short term to log the forest in an unsustainable way. Although the need for sustainable harvesting of timber trees is mentioned frequently in the context of Amazonian development, it is hardly practiced, at least on any significant scale.

Figure 3.26. A rubber tapper's hut nestled among the buttress roots of a kapok tree. The fire was used to help coagulate the latex. Smoked rubber, stacked like pancakes, can be seen by the paddle at the base of the tree. From Brown, C. B., and W. Lidstone, *Fifteen thousand miles on the Amazon and its tributaries* (Edward Stanford, London, 1878), p. 444.

Farmers in both upland areas and on the floodplain are incorporating timber species in their agroforestry plots, and in the future plantation forestry is likely to emerge as a major land use in the region.

Floodplain Forests Falling through the Cracks

Undervalued resources run the risk of being ignored, and eventually lost. Floodplain forests are victims of market imperfections. In spite of their many uses, floodplain forests of the Amazon have been largely undervalued for a variety of reasons. First, they were thought to occupy a marginal environment, subject to extreme stresses, and thus contain little of value for economic development. Second, once most of the valuable timber had been logged off the middle stretch of the Amazon, it was thought that conversion to crops or pasture would be a logical, even desirable process. Third, the critical role that floodplain forests play in the food chain of numerous fish important for commerce and subsistence was not understood in detail until Michael Goulding began his pioneering work on the ecology of Amazonian fishes in the mid-1970s.[10]

The consequences could be disastrous. And still, some agronomists have called for cutting down the "useless" forests of the Amazon floodplain and draining the land for the cultivation of annual crops.[11] Linwood Pendelton of the Kennedy School of Government at Harvard asserts that the vocation for fertile alluvial soils in the humid tropics is for agriculture, rather than harvesting of forest products.[12] Floodplain forests, it is argued, have fewer species than upland forests, and therefore should act as a safety-valve for settlement and food production in order to "save" the biodiversity of upland forests. But the biodiversity of forests on the floodplain is remarkably high, in part because of the shifting availability of habitats and differences in micro-relief. While the Amazon floodplain can certainly play a greater role as a breadbasket for South America, the notion that the *várzea* forests are essentially dispensable is fraught with dangers. Floodplain forests could be a major asset for diversifying agriculture on the floodplain, rather than an obstacle to crop production.

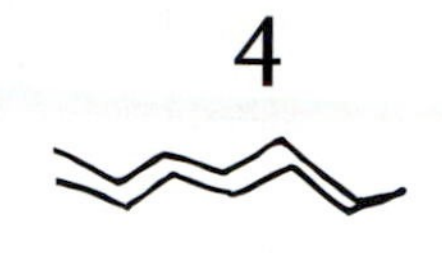

The Emergence and Impact of Livestock Raising

Apart from the Muscovy duck and some stingless bees, no truly domesticated animals appear to have been kept by indigenous peoples in Amazonia. Abundant sources of animal protein along the rivers, especially along muddy rivers such as the Amazon, may have dissuaded groups from taking the trouble to tame, house, and feed animals. Many mammals and birds were kept as pets, but these were usually captured in the wild while still young. Some "pets" were eventually eaten, but others, particularly parrots, macaws, and harpy eagles, were kept to supply feathers for ornamental wear. Several Indian groups once kept thousands of giant river turtles in aquatic corrals along the Amazon, but these "river cattle" were captured after laying eggs on beaches at low water, rather than raised in captivity.

Indigenous people thus did not transform Amazonian landscapes to raise animal protein. Cattle, water buffalo, horses, donkeys, pigs, sheep, and goats were all introduced to the region after contact with Europeans. During the colonial period, Europeans began cutting small patches of forest on uplands near rivers to expand pasture for cattle and horses, but such clearings were limited because herd size was generally small and labor scarce. Large-scale clearing of upland forests in the early colonial period would not have led to good pasture anyway, because African grasses were not imported to Brazil until the eighteenth century. For several hundred years after Europeans first arrived in Amazonia, cattle were confined mostly to Marajó Island, the savannas of Roraima, and the Santarém area, where upland savannas were available to tide the animals over until the low-water season, when they could graze again on the lush *campos* of the Amazon floodplain.

The rapid growth of towns and cities, particularly in the last few decades, is driving a dramatic expansion of cattle ranching throughout Amazonia. Landscapes along the Amazon, particularly the middle stretches, have changed accordingly. Ranchers deforest the floodplain, even backswamp forest, as at Murumuru near Santarém, to encourage native grasses. Unlike in the past when indigenous groups made small clearings for crop production, most of which eventually reverted to forest, ranchers are opening the landscape on an unprecedented scale. Now vast stretches of woodland are being taken out, with little likelihood of them reverting to forest at least in the foreseeable future.

Introduction of Cattle

One of the myths of Amazonian development is that cattle ranching emerged as a significant economic activity only in the twentieth century, particularly since the 1960s. But the activity has been a major land use along the Amazon floodplain from roughly Faro downstream to Prainha since at least the mid-nineteenth century. And it has always been the major economic activity on Marajó since the arrival of Europeans. Furthermore, ranching was a major, if not the single most important economic activity on the Amazon floodplain in the vicinity of Monte Alegre and Óbidos over 120 years ago. In 1878, a British engineer on his way to the now defunct Madeira-Mamoré railroad described the area of Monte Alegre:

> Here there are large pampas, on which cattle are reared in great numbers. These "fazendas de ganado" or cattle-runs, are very valuable properties, especially those that have hilly lands on them, where the cattle can take refuge during the periodical inundations.[1]

In the mid-nineteenth century, cattle raising undoubtedly increased the population of vultures, particularly the black vulture. At Prainha along the lower Amazon, for example, the processing of fish and the butchering of cattle and turtles attracted large flocks of vultures in the last century. What has changed in the latter part of the twentieth century is the scale of cattle ranching; a virtual stampede of cattle and water buffalo has triggered large-scale deforestation and has stirred conflicts with crop farmers.

The greatly increased herds of cattle and water buffalo on the Amazon floodplain have wrought havoc on the forests, including waterlogged jungles of arum, the fruits of which are used for turtle bait. Large stands of arum, known as *aningal* in the Brazilian Amazon, have succumbed to repeated burning to improve pasture. Green-stemmed arum, which can reach three or four meters, was once common on the floodplain, serving as a refuge for many species of wildlife. Along Paraná do Orive on Ilha do Carmo, for example, residents remember miniature forests of arum behind their houses; now open grassland with only pockets of woodland dominates the view from their windows. Near Monte Alegre, an island of arum in a large lake serves as heronery; the white wings of the several species of heron contrast spectacularly with the large, succulent arum leaves. Such sites are increasingly rare, with repercussions also on turtle populations. Both the giant river turtle and the yellow-spotted Amazon turtle are fond of arum fruits, and the diminished stands of the aquatic aroid has undercut an important food resource for those avidly sought game species.

Cattle were brought to the New World when Columbus landed at Hispaniola on his second voyage in 1494, but when they reached Amazonia is unclear. They were first imported to Brazil in the 1530s to several locations, including the mouth of the São Francisco River in the Northeast and São Vicente, near the present-day city of Santos in São Paulo. Most of the economic activity and limited settlement by Portuguese during the colonial period occurred along the northeast and southeast coast of Brazil. Amazonia was largely a backwater, and cattle may not have reached the basin until 1644, when they were brought to Belém twenty-eight years after the city was established.

Ranches were established on the seasonally wet grasslands of nearby Marajó Island in 1680, and by 1806, 226 ranches were operating on Marajó with 500,000 head of cattle. Catholic missionary orders, particularly the Mercedarians, Jesuits, and Carmelites, took the lead in establishing cattle herds on Marajó and in several other parts of the basin. The Switzerland-size island soon emerged as the main cattle ranching area for eastern Amazonia (figure 4.1), retaining that position until the 1960s when bulldozers pushed corridors through upland forests and settlers followed in their wake.

In the late 1700s, cattle were released on the broad savannas of Roraima, which cover 20 million hectares in the northern part of the Amazon basin, to supply Manaus with meat. Few herds were kept on the Amazon floodplain near Manaus in the eighteenth and nineteenth centuries. Careiro Island near Manaus, a dairying center for much of the twentieth century and now a significant source of beef for the city, appears to have had few ranches in the mid-nineteenth century; cattle raising there began in earnest only in the 1880s. Until the late nineteenth century, few cattle were kept on the Amazon floodplain above Faro. One of the largest

Figure 4.1. A cowboy (*vaqueiro* in Brazil) on Marajó Island, a traditional cattle ranching area at the mouth of the Amazon. "Vaqueiro de Marajó" drawing by Percy Lau, in *Tipos e Aspectos do Brasil*, (Fundação Instituto Brasileiro de Geografia e Estatística, Rio de Janeiro, 1975), p. 67.

herds was kept at Tefé, where an officer in the Royal Navy, Henry Maw, recorded some two hundred cattle in 1828.[2] Other communities along the Upper Amazon had few if any cattle. The Peruvian Amazon seems to have been completely devoid of cattle during the colonial period.

The introduction of zebu cattle from India in the 1870s, especially the Nelore breed, provided a boost to cattle production in Amazonia, including the floodplain. Zebu cattle tolerate heat and humidity better than the humpless *criollo* cattle brought in earlier from Iberia. Furthermore, tall Zebu can wade up to their necks to feed on floating grasses, whereas relatively short-legged European cattle are less well adapted to floods (figure 4.2).

Pasture Formation

The process of pasture formation on the floodplain is different from that on the uplands. On *terra firme*, the forest is cut and burned in the dry season, typically cultivated for a year or two with such crops as rice, maize, or manioc, and then sown to grasses of African origin. On the floodplain, the trees are felled and burned and then nature takes its course. Native grasses generally take over, along with other volunteer plants. In effect, ranchers and farmers are expanding the area of "natural" campos.

Fire has long been used to expand floodplain meadows. People have been deliberately setting fire to the *campos* along the lower Amazon for more than three centuries in order to improve pasture for cattle at low water. It seems likely that fires were also set on the meadows during pre-contact times, just as indigenous groups deliberately torched upland savannas along some stretches of the Amazon to facilitate hunting. On Mexiana Island north of Marajó, for example, fires

Figure 4.2. Zebu cattle feeding on canarana grass, brought by canoe to a home garden on the floodplain. Zebu is a cattle breed that arose in India and was brought to Brazil in the 1870s. Carariacá, Pará, May 1996.

were set annually in the mid-nineteenth century to promote fresh growth on the seasonally flooded grassland. Likewise, ranchers in the vicinity of Óbidos and Monte Alegre habitually burned floodplain meadows during the dry season just before the floodwaters began to rise.

Such fires were sometimes so intense, especially during dry years, that cacao orchards were destroyed, as in the vicinity of Óbidos during the last century. Out-of-control fires probably penetrated the perimeter of surrounding floodplain forest during severe dry seasons, thereby pushing back the woods. In this manner, trees were undoubtedly prevented from colonizing grassy areas that would eventually have become forest.

More than a century ago, an American naturalist recognized the role of fire in shaping floodplain grasslands along the middle Amazon:

> Near the main channels the meadows are much broken by these bushes and swamps; but far back, and sheltered in bays of the highland, they are as level and clean as a wheat-field, bright velvety green, rippling with the wind like a great lake. Everywhere the grass is dotted with cattle. Such places, indeed, owe their beauty to the yearly fires with which the herdsmen cleanse their surface.[3]

Floodplain meadows are also set ablaze at low water today, sometimes inadvertently. A Brazilian anthropologist, Lourdes Furtado, reports that when fisherfolk cook their meals along river banks near Óbidos, the surrounding grass sometimes catches fire, especially if the dry season is severe.[4]

Some farmers and ranchers also sow indigenous grasses on a limited scale to foster pasture on the floodplain. At Fazenda Bom Retiro along the margins of Lake Paracari, a small-scale rancher has been steadily clearing forest to encourage spontaneous black-bellied tree duck grass for his water buffalo herd. As suggested by the Brazilian name for this native grass, *capim marreca* (black-bellied tree duck grass), wild duck are also apparently fond of the grass, both for cover and for foraging. In the vicinity of Carariacá near Santarém, one farmer who also raises cattle collects seed of canarana lisa grass to sow on mud flats as the floodwaters recede. The "slippery" (*lisa*) canarana grass is much appreciated by cattle. On Careiro Island, the Brazilian geographer Hilgard Sternberg noted that some ranchers were planting canarana to increase cattle fodder in the early 1950s.[5]

On Ilha Grande near Óbidos, farmers plant Pará grass on high banks as the flood withdraws, also for cattle fodder, and a large rancher near Santarém is planting cuttings of the stoloniferous grass on part of his floodplain property to improve forage, particularly for calves and horses. In spite of its name, Pará grass is native to West Africa; it was introduced to Brazil in the early nineteenth century as bedding in slave ships. Pará grass, called mojuí or colônia along the middle Amazon, is the only non-native grass planted on the Amazon floodplain. In contrast, upland pastures are invariably planted to African grasses, particularly guinea grass and brachiarão.

Adaptations to the Flood Cycle

When the rising waters become too deep, ranchers and farmers are faced with four options for their cattle: confine them to a platform (*maromba*); pen them in a small corral (*caiçara*) at the water's edge; locate a hill on the floodplain; or transfer the cattle to an upland pasture or savanna. Larger ranchers always remove their cattle to upland pastures. Small-scale operators usually opt for floating or elevated platforms, corrals by the waters' edge, or if they have sufficient resources, rented pasture on the uplands. A century ago, upland pastures were rare; now most of the cattle along the Amazon spend the high-water season on such grazings carved out of the forest.

Not all options are available to a given farmer or rancher. Only a few areas have floodplain hills where cattle can be kept at high water. One such area is along the Paraná do Ramos near Maués in Amazonas. The hills are outliers of adjacent uplands that were isolated during various periods in the Pleistocene when sea level dropped and watercourses scoured their beds deeper. When the ice caps melted, sea level rose, river currents slackened and sediment was dumped, thereby filling in gaps around the hills. Such hills can be easily recognized because they often support Brazil nut trees, as near the village of Matá near Óbidos. And in the vicinity of Alenquer, a mound of *terra firme* on the floodplain contains fragments of pottery and discarded bivalve shells, suggesting that such year-round dry spots on the floodplain were also in use long ago.

Marombas for keeping cattle on the floodplain at high water come in two forms: floating corrals (figure 4.3) and—less common—platforms on poles. On Careiro Island in the 1950s, platforms on poles contained up to 180 cattle. Once ubiquitous, floating corrals are increasingly rare as more farmers and ranchers opt to rent or buy upland pasture, thereby accelerating deforestation on non-flooded areas as well (figure 4.4). The most commonly used tree to construct floating cor-

Figure 4.3. A floating corral (*maromba*) containing about twenty cattle standing and resting on assacu logs. The cattle have been confined to the raft for several months. The *maromba* is in an area formerly occupied by forest, as evidenced by the tree stump to the right. This bank of the Amazon has been through various cropping cycles from cacao, to jute, and now pasture. Ilha Grande near Óbidos, Pará, June 1994.

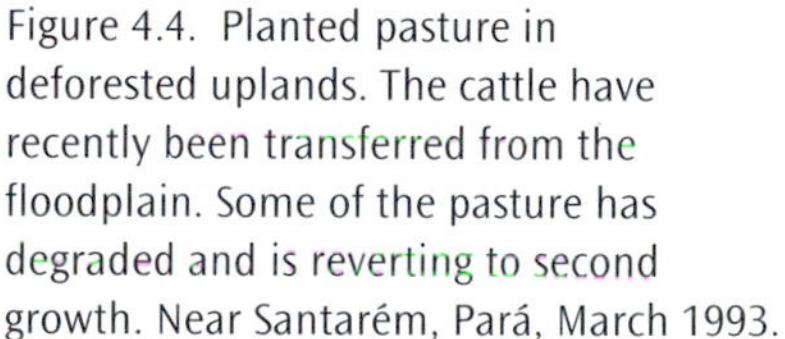
Figure 4.4. Planted pasture in deforested uplands. The cattle have recently been transferred from the floodplain. Some of the pasture has degraded and is reverting to second growth. Near Santarém, Pará, March 1993.

rals is assacu. The massive logs are intercepted as they float downstream at the onset of the floods, and are lashed together with vines before the waters cover the bank. In the past, equally impressive logs of the kapok tree were also used to assemble marombas, but excessive logging has drastically thinned the trees. Assacu still survives in reasonable numbers because it is not useful for timber; locals, however, employ the bark for medicinal purposes. Along the middle Amazon, ten to forty cattle are usually placed on marombas in February as the waters rise. In late June and early July, as the flood gradually recedes, they are allowed out to forage for brief periods in knee-deep water (figure 4.5). By August, the rotting maromba is abandoned as the cattle can now graze on their own.

Cattle in both marombas and caiçaras are stall-fed a variety of grasses (figure 4.6). This twice-daily practice has been carried out along the middle Amazon since at least the mid-nineteenth century. Canoes laden with floating grasses, such as pemembeca and canarana, are a common sight on the *várzea* soon after dawn and in the late afternoon (figure 4.7). Stands of canarana are among the most productive natural vegetation communities on earth. The lanky grass grows stems as long as eight meters to keep pace with the rising waters; while still anchored to the bottom, it develops floating adventitious roots at the nodes. Pemembeca has a different strategy: as floodwaters continue to swell, the stems eventually

Figure 4.5. Cattle feeding on cut canarana grass. The animals, which belong to a small farmer whose house is nestled among fruit trees on the left, will shortly be herded back on to the floating corral (*maromba*) where they have spent the last four months. Ilha Grande near Óbidos, Pará, June 1994.

Figure 4.6. Cattle heading for their afternoon feed of cut pemembeca grass. These animals belong to a relatively prosperous small farmer and are being kept in a corral (*caiçara*) along the floodplain margin during high water. The cattle are in poor condition, which will be improved only when they are transferred to floodplain pastures in about six weeks. Carariacá, Pará, June 1994.

Figure 4.7. A small-scale rancher returning shortly after sunrise with pemembeca grass cut on the floodplain. The grass will be fed to a small herd of water buffalo. Although the floodwaters have been receding for a couple of months, there is still not enough pasture on the ranch to sustain the herd without supplemental feeding. Lago Paracari near Santarém, Pará, August 1993.

separate and currents carry away mats of the grass with their floating roots. Clumps of the floating grasses are hacked off with machetes and loaded onto canoes.

Another floodplain plant occasionally fed to cattle is patazana, a member of the Marantaceae, a family better known to North Americans and Europeans for a diverse array of indoor ornamental plants with attractive leaves. Also known as pariri, it grows in rank profusion in open habitats as the floodplain waters recede. In the vicinity of Arapixuna, locals cut the two-meter-tall gladiolus-like plant to provide feed for cattle during their boat ride from uplands to seasonal grazing grounds on the floodplain.

The high-water season is always stressful for cattle. Upland pastures are often weed-infested, especially if used only seasonally, and cattle are typically emaciated by the time they are brought to the floodplain. Nutritional diseases therefore take their toll by the time water levels are falling. For cattle kept on marombas, the cramped conditions over several months can lead to injury and premature death, such as when legs become wedged between logs and break. Farmers sometimes place a trough in the maromba containing a mixture of salt and cobalt as a nutritional supplement, but cattle still lose condition. The emergence of floodplain pastures is a joyful event for livestock and their owners alike.

Some farmers living along the interface between uplands and the floodplain employ a combination of planted pastures on *terra firme* and caiçaras along the water's edge. At Carariacá near Santarém, for example, one farmer maintains a herd of 120 cattle with only 8 hectares of upland pasture. At high water, the cattle are released in the morning to graze on the weed-choked upland pasture. By noon the cattle are thirsty and return to their pen at the floodplain edge to drink and rest. In the late afternoon, the farmer's sons and hired workers fetch floating grasses to supplement the cattle's feed. At low water, the herd spends several months on the floodplain without supplemental feeding, and the upland pasture rests. Without access to the floodplain, the farmer would need about 120 hectares of upland pasture to support his herd.

Today, because most people who keep cattle on the floodplain take them to upland pastures during the high-water period, the increased herd size has triggered more deforestation on uplands as well as on the floodplain. When ranchers have access to the Amazon floodplain, stocking rates on upland pastures are much greater than for those who must maintain cattle year-round on upland pasture (table 4.1).

A farmer who lives in Aninduba, overlooking the Amazon upstream from Santarém, owns 30 cattle but has cleared only 6 hectares of upland pasture. Across the river at Flexal in the Municipality of Óbidos, another farmer keeps 15 head on only 3.5 hectares of upland pasture at the back of his house. Because the upland pastures are grazed only four to six months of the year, they can support a stocking rate of roughly 2 to 6 head per hectare. In areas where floodplain grasses

Table 4.1. Stocking rates for cattle on upland pastures for the 4 to 6 months that grazing areas on the Amazon floodplain are under water

Upland pasture (ha)	Herd size	Stocking rate (head/ha)	Location in Pará
8.0	120	15.0	Carariacá near Santarém (upland pasture inland from the village)*
45.0	20	6.2	Fazenda Pensamento, Ilha de São Miguel, near Surubim-Açu (upland pasture near Monte Alegre)
40.0	230	5.7	Paraná do Baixo near Óbidos (upland pasture at Fazenda Bom Futuro, km 33 of Óbidos–Monte Alegre highway)
6.0	30	5.0	Aninduba (upland pasture on bluff overlooking Amazon)
3.5	15	4.3	Flexal near Óbidos (upland pasture inland from lake shore)
30.0	70	2.3	Ramal do Andirobalzinho, a side road from the Santarém–Alter do Chão highway (floodplain pasture is rented)
12.0	20	1.7	Arapixuna (upland pasture on bluff overlooking Amazon floodplain)
280	270	1.0	Fazenda São Lourenço, Paraná Miri, near Alenquer (upland pasture at Fazenda Morado do Sol, 23 km inland from Alenquer)**

*Cattle provided supplemental feed using floating grasses from the Amazon floodplain.
** 180 cattle and 90 water buffalo (35 sheep not included in stocking rate).
Stocking rates are only approximate because some cattle may be bought or sold during the time the herd is kept on upland pasture. Also, small flocks of sheep, herds of goats, and work animals such as horses and mules may share pastures.

are abundant and can be easily brought to the water's edge, upland pastures can support 15 head of cattle per hectare for a few months, much higher than the regional average of 1 head per hectare for year-round grazing.

In the vicinity of Monte Alegre and Santarém, ranchers and some farmers graze their cattle on woody savannas at high water. Free-range cattle are put out from roughly March to August, a period spanning the latter part of the rainy season and the early part of the dry period. Grazing is thus at its best. As the savanna grasses are consumed or begin to wilt under the searing heat of the dry season, the animals are rounded up and are driven or taken by boat to emerging floodplain meadows. The advantage of these savannas is that they require no maintenance other than periodic burning when the rains subside. Planted pastures, on the other hand, require weeding every year, otherwise they will be taken over by second growth and re-sprouting tree stumps. The carrying capacity of savannas is generally lower than planted pasture; the former supports only one head per three to five hectares in the Santarém area.

Planted pastures on the uplands, however, are increasingly favored for grazing during the high-water period. Several reasons account for this trend—the network of roads built since the 1960s provides better access to upland forest; savannas are not always convenient; cattle on smaller upland ranches can be more easily corralled at night to protect them from rustlers than free-range cattle on savannas; and the carrying capacity of *terra firme* pastures, although modest, is superior to that of savannas. Finally, the greatly increased herds cannot be easily accommodated on the savannas close to the Amazon River.

A majority of the large-scale ranchers own property on the uplands and floodplain. In some cases, the properties are separated by more than one hundred kilometers. In September 1997, for example, I encountered a herd of some eighty cattle at km 22 of the Santarém to Curuá-Una highway; the rancher and his *vaqueiros* (cowboys) had been on the road for about a week making their way from upland pasture in the headwaters of the Moju River about 120 kilometers from Santarém. After they reached the riverside town, the emaciated herd would still face a boat journey of several hours to the owner's floodplain ranch near Alenquer. Smaller-scale operators usually rent pasture on *terra firme*. In 1993, farmers and small-scale ranchers near Santarém were paying monthly between 50¢ and $1.10 per head to rent upland pastures.

Conversely, some smaller ranchers on the uplands rent floodplain pasture at low water. The experience of one rancher from the northeastern state of Ceará illustrates the tradeoffs and advantages of cattle transhumance. He owns 280 hectares along Ramal do Andirobalzinho, a side-road from the Santarém to Alter do Chão highway, and keeps seventy head of cattle on 30 hectares of pasture. This relatively high stocking rate for the uplands is only possible because the cattle are taken to the Amazon floodplain, some thirty kilometers away, during low water. Without this transfer option, he would have to clear more forest to accommodate his cattle. Of the 280 hectares on his property, 100 hectares is in forest, while second growth has reclaimed 145 hectares. Three hectares a year are cleared to plant rice, maize, beans, and pineapple. Cattle raising requires much more space than crops, and deforestation of the property would undoubtedly accelerate if prices for pineapple or other farm produce declined, or if floodplain pastures were no longer available.

Landowners on the Amazon floodplain generate modest income from renting pastures. Typically, they allow others' livestock to mingle with their own for a small charge or in-kind contribution. In 1992, ranchers and farmers along Ramal do Andirobalzinho were paying two dollars per head to graze their cattle on floodplain grassland for the entire low-water season. In the Óbidos area, the prevailing rent for floodplain pasture was about a dollar and a half per head in 1996. In the Santarém area, rental rates for floodplain pasture are about one-fifth those charged for placing cattle on upland pasture, a reflection of the lower operating costs on the floodplain.

For farmers and ranchers living along Ramal do Andirobalzinho near Santarém, the cost of transporting cattle to and from floodplain pastures, some thirty kilometers away, was four dollars per head in 1992, including truck rental for the upland portion of the journey. At Flexal in 1996, transportation costs were about two dollars per head for a five-kilometer boat ride to suitable floodplain pasture.

As late as the 1940s, steam-powered boats transported cattle on the Amazon floodplain; they have since been replaced by diesel-powered craft. The last steam-powered boat in the Santarém area was hauling jute before it was taken out of service around 1950. The holds of many two-story boats are converted so that they can carry cattle to and from the *várzea*; the remainder of the year they carry cargo and passengers. Single-story boats are also used to convey cattle, while large herds are hauled on barges equipped with stalls.

The cost of transporting cattle and renting floodplain pasture is compensated by their weight gain under the typically superior grazing conditions of the floodplain. During low water, cattle are especially fond of such naturally occurring grasses as long-stemmed canarana, pancuã, black-bellied tree duck grass, wild rice, and taboka, a bamboo-like grass. Cattle only resort to murí grass when more palatable grasses are not available.

Overgrazing can occur as the waters begin to recede and cattle are placed on the floodplain before sufficient pasture has formed on the emerging land. On the whole, though, overgrazing does not appear to be a major problem on the Amazon floodplain, at least with respect to triggering environmental damage such as accelerated soil erosion.

The Stampede toward Water Buffalo

Water buffalo are relative newcomers to the Amazon floodplain, but their numbers are increasing in proportion to cattle. Water buffalo were first brought to Brazil in the late nineteenth century and today most of the herds are in the Amazon, especially on Marajó Island. Over the last several decades, the hardy animals, raised for meat and to a lesser extent for cheese, have been moving upstream. Although a few water buffalo were being raised near Santarém in the early 1900s, they did not make their appearance in any significant numbers along the middle Amazon until the 1950s. Murrah and Mediterranean (*Mediterrâneo*) are the preferred breeds of water buffalo on the Amazon floodplain. Few ranchers in the Santarém area keep the more aggressive Carabou breed.

Felisberto Camargo, director of the Agronomic Institute of the North (IAN; Instituto Agronomico do Norte) in Belém, was responsible for introducing water buffalo to the Santarém area. Now part of the national agricultural research program (EMBRAPA; Empresa Brasileira de Pesquisa Agropecuária), IAN also brought some individuals of the Murrah breed of water buffalo to Oriximiná for an auction in the 1950s, thereby sparking interest in the exotic animal among some cattle ranchers. Along other stretches of the middle Amazon, such as Manaus, entrepreneurs were responsible for introducing water buffalo over the last few decades. The current population of the animals in the vicinity of Nhamundá, probably numbering several thousand, can be traced to five individuals brought from Marajó Island by a local rancher, Pedro Baranda.

During the 1980s, water buffalo herds appear to have started taking off along the middle Amazon. While in 1977 I saw only one small herd near Itacoatiara, Amazonas, and only 4,000 water buffalo were recorded for all of Amazonas State in 1975, by the early 1990s, the Itacoatiara area alone probably had several thousand water buffalo. By one estimate, some 800,000 water buffalo were in the Brazilian Amazon by the late 1980s. Marajó Island alone now has close to a half a million head. In 1990, the Jari pulp operation was raising a herd of 13,000 on the Amazon floodplain portion of its property, which covers more than 1 million hectares.

Originally the domain of large landholders, water buffalo are now being adopted by small-scale ranchers and some of the more prosperous farmers. The increased size of the herds is enriching ranchers, but is also exacting a heavy toll on crops of small farmers. Water buffalo are better adapted to flooding than cattle and can linger on the floodplain after cattle have been rounded up and taken to uplands or placed in marombas (figure 4.8). In some areas on the middle Amazon floodplain, they are kept year-round, reducing transportation costs for ranchers

Figure 4.8. Water buffalo thrive along the Amazon because they can stay on the floodplain longer than cattle. Virtually all of the dry land has disappeared in this area due to rising waters, but water buffalo will probably linger for another month before being transferred to upland pasture. Igarapé-Açu near Santarém, Pará, March 1993.

and eliminating the need to rent or purchase upland pasture. To a certain degree, water buffalo are taking over some of the niche left by the greatly reduced stocks of manatee and capybara, which also graze aquatic plants.

The meat of water buffalo is usually sold as "beef" in markets, and few individuals are able to tell the difference. A native of Marajó claims he can tell the meat of water buffalo by its white fat; beef, he says, has yellower fat. Such distinctions are unlikely to hold up under all conditions, however. Water buffalo from the Santarém area are shipped to markets as far afield as Belém, Macapá, and Manaus. About 84 percent of those slaughtered in Manaus come from Pará, but the contribution from Amazonas state is increasing as herds expand, particularly around the confluence of the Amazon and Madeira rivers and in the vicinity of Parintins.

Ranchers near Santarém and Alenquer cite a number of reasons why water buffalo are increasing on the floodplain. Water buffalo put on weight faster than cattle, reaching an average of 350 kilograms by their third year, in comparison to 250–300 kilograms for cattle. Although water buffalo have a longer gestation period than cattle (ten months versus nine), they reach sexual maturity when they are two years old, a year earlier than cattle; thus herds build up faster. Water buffalo are also hardier than cattle, require less veterinary care, and can get by on poorer grazing conditions.

Water buffalo also yield more milk, producing an average of five liters of milk a day along the Amazon compared to about three liters for cattle. The Murrah breed has proved the best milk producer in the Santarém area. Furthermore, their exceptionally rich milk is preferred for making mozzarella cheese, which is in demand in urban areas with the growing popularity of pizza. Also, cheeseburgers in Brazil are made with white cheese, preferably mozzarella, in contrast to orange-colored American cheese preferred by fast-food addicts in the United States. Although much of the mozzarella consumed in Brazil and elsewhere is made from cattle's milk, a better quality product is obtained from water buffalo. Finally, water buffalo are more docile and easier to handle than cattle. On Marajó, *vaqueiros* even ride on sturdy water buffalo.

Driving Forces

Rapidly growing urban markets are the main force propelling the expansion of cattle and water buffalo ranching along the Amazon and other parts of the region. In the middle of the last century, only about thirty head of cattle were slaughtered per day to feed Belém, hardly a big enough market to encourage large-scale clearing of forest for pasture. Even by the early 1920s, only about one hundred cattle and water buffalo were butchered daily for the Belém market. Today, at least twelve hundred cattle and water buffalo are processed every day in Belém to meet local demand. Hard figures on the slaughter rate for cattle in Belém are difficult to obtain because at least a third of the cattle are butchered in clandestine operations.

A similar situation prevails in Manaus, another large market for beef, with much beef coming from "irregular" sources. For much of the nineteenth century, a few dozen cattle were taken yearly by boat to Manaus from the savannas of Roraima, the city's main supplier of meat. Today Manaus consumes almost that many cattle and water buffalo in an hour, mostly from herds that are raised on the Amazon floodplain and upland pastures. Roraima no longer sends appreciable numbers of cattle to Manaus, even though a road is now open linking the state capitals. In fact, Boa Vista, the capital of Roraima, has ceased being self-sufficient in meat and has to import beef. And Manaus is reaching as far afield as Argentina to satisfy its growing demand for beef.

Cattle and water buffalo ranching is expanding because the animals are highly liquid assets, are considered prestigious, serve as a secure form of savings, and require minimal labor. At the São Lourenço Ranch on the floodplain near Alenquer, for example, the largely absentee landlord hires only two men to look after his 180 cattle, 90 water buffalo, and 15 sheep. Labor requirements at the Pensamento Ranch on São Miguel Island some thirty kilometers downstream are similar: two employees look after 280 cattle, 44 water buffalo, and 12 sheep. The São Bento Ranch along the main bank of the Amazon at Costa do Pindurí near Santarém gets by with one caretaker and his two sons, aged eleven and thirteen years, to look after 96 water buffalo and 50 sheep.

Although often criticized as inappropriate for Amazonia, the raising of cattle is an integral part of many small farms and helps sustain them. Although ranching does not generate much employment, the ability to quickly liquidate a few cattle has saved many a farmer from having to sell his land to cover an emergency. And as will be noted in the chapter on annual crops, livestock provide valuable manure for crops, especially vegetables.

Fiscal incentives, often cited as a major reason for the "grass rush" in Amazonia, never promoted cattle ranching on the Amazon floodplain as they did in some upland areas in the 1960s and 1970s. In Brazil, the regional development agency (SUDAM; Superintendência do Desenvolvimento da Amazônia) has provided tax breaks to some entrepreneurs to start water buffalo herds or to expand their existing herds. For example, twenty-seven water buffalo projects were approved for the state of Amapá between 1980 and 1986. But such incentives provided a modest catalyst, rather than the main force behind the rapid spread of water buffalo herds during the 1980s and 1990s.

The removal of all fiscal incentives for cattle ranching by 1991 has not slowed the expansion of pasture in upland areas. The cost of credit is so high that few ranchers risk their properties as collateral for bank loans. A common remark about bank loans for raising livestock or crops is that prevailing interest rates would "undercut their operations" (*"inviabiliza tudo"*).

The main market is for meat, mostly consumed within Amazonia. Indeed, the region is barely self-sufficient in beef and sometimes resorts to bringing in cattle or meat from other parts of Brazil or abroad. The aversion of some consumers in North America and Europe to eating "red" meat for health reasons has not reached Amazonia. Even the poor attempt to buy meat, albeit the cheaper cuts, at least for Sunday. In the Itacoatiara area, I have seen fishermen sell part of their catch in order to buy meat for the family. The prestige associated with cattle raising—and putting beef on the table—in Iberia was transferred in the early colonial period and remains firmly rooted in the regional culture.

Milk and cheese production, a low-water activity along the river, generates modest revenues for some farmers and ranchers. Traditionally, milk consumption has been modest in the region, but demand for the product is increasing with the swelling ranks of the middle and upper classes in cities. The influx of people from central and southern Brazil, who are accustomed to milk and other dairy products, is also improving market prospects for milk production in the Amazon.

Still, much of the milk consumed in urban areas of the Amazon is reconstituted from powdered milk produced in southern Brazil or western Europe. In Manaus, for example, more than 90 percent of the milk consumed is prepared at home from powdered milk. And even the "fresh" milk marketed by the single dairy in Manaus is mostly reconstituted from powder. Some urban residents explained to me that they grew up accustomed to the flavor and texture of powdered milk, and prefer to use it for their morning coffee even when fresh milk is available. One mother explained that powdered milk beaten with a whisk produces a creamy froth, whereas regular milk is "flat."

Despite the regional predilection for powdered milk, milk production can be a worthwhile undertaking for farmers and ranchers within a two or three hours' boat ride from a major town or city. Typically, cows are milked before dawn, the milk cans are then loaded onto scheduled boats carrying passengers around daybreak, and arrive in town before the milk spoils. Distributors then deliver the unpasteurized milk to customers, who boil it before consumption. Milk is primarily destined for children, although some adults add it to their morning coffee. It is rare, however, to see an adult drink a glass of milk after breakfast in urban areas of the Brazilian Amazon. Only the largest cities along the Amazon have dairies where the milk is sterilized and refrigerated. Amazonmilk, the only dairy in Santarém, the third largest city in the Amazon, closed in 1995 after only two years of operation due to management and marketing problems. With the capacity to process 30,000 liters of milk a day, the dairy had been functioning at only a fraction of its capacity before folding, mostly buying milk from upland and floodplain producers. For those farmers and ranchers further removed from sizable towns and cities, highly perishable milk is processed into cheese, curdled milk (*coalhada*), or fudge (*doce de leite*) before shipment. Cheese making is a cottage industry for small farmers as well as some of the larger ranchers. The fatty, ranch cheese produced on the floodplain can be stored for several weeks at ambient temperatures without spoiling. Cheese produced from water buffalo milk is especially high in calories because the milk contains 8 percent butterfat, compared to about 3 to 4 percent for cattle. Relatively bland by U.S. and European tastes, the somewhat greasy and rubbery cheese produced on the Amazon floodplain finds a ready market in towns and cities along the Amazon. Nutritious whey, a byproduct of cheese making, is fed to pigs. The curdled milk produced along the Amazon has the consistency of yogurt, but without the live cultures, and can be stored for a day or two before it spoils.

Ecological and Cultural Impacts of Ranching

The expansion of cattle and water buffalo herds has come at the expense of crops as well as forest, and has therefore generated conflicts between farmers and owners of cattle. When cattle or water buffalo are present, crops can be grown successfully only when they are fenced. The problem is that few floodplain ranchers fence their properties, and small farmers can afford to do so only for small vegetable plots near their homes.

This is an old issue. A century and a half ago, Henry Walter Bates noted that *ribeirinhos* on the Amazon floodplain at Maicá near Santarém and in the vicinity of Tefé could not grow crops because of undisciplined management of cattle.[6] Some farmers in the vicinity of Tefé would even leave out bowls of juice squeezed from bitter manioc to poison cattle that had been troubling their gardens. But even fences will not keep hungry water buffalo away from succulent vegetables or maize. Water buffalo can easily swim around fences. And unless the fences are solid and well made, both cattle and water buffalo will breach them. Ranchers are supposed to compensate farmers for any crop damage caused by free-range cattle and water buffalo, but rarely do.

The Amazon floodplain will only realize its potential as the next great agricultural frontier of Brazil when cattle and water buffalo are better managed. When ranching abuts crop farming, the latter usually loses out. And it is not simply a case of absentee owners of large ranches causing all the damage. Small farmers are also contributing to the problem as they adopt large livestock, or expand their existing herds. Farmers who live in Arapixuna, for example, no longer plant crops in front of the village because of the growing number of cattle acquired in recent years by some residents. A similar story is told in other parts of the floodplain of the middle Amazon, such as at Amador near Óbidos. When hemmed in by cattle, farmers are typically forced to establish their fields further afield on the Amazon floodplain, or sell out.

In many parts of Southeast Asia, water buffalo are an integral part of small farms, where they prepare rice fields for planting and haul agricultural products to market. Water buffalo and farmers are in harmony along the great Mekong River, which is far more densely settled than the Amazon floodplain. But the animals are not allowed to wander freely; rather they are tethered to keep them out of fields and are stall-fed at night from grasses cut from roadsides and the edges of fields. Water buffalo and cattle can continue to make a valuable contribution to the regional economy of Amazonia, but management practices will have to change to reduce crop destruction and environmental damage.

The trajectory of water buffalo ranching in the Peruvian Amazon has been markedly different from that prevailing in Brazil. In 1981, the regional government of Loreto in the Peruvian Amazon set up low-cost loans to encourage water buffalo ranching along the Ucayali, a major headwater river of the Amazon. Within a few years, several thousand head were imported from Brazil. Loan agreements stipulated that the animals were to be corralled to prevent crop damage. But few pens were built and conflicts soon arose between farmers and ranchers. Problems with water buffalo persisted and the loan program was disbanded. Pressure from rural unions has also forced some livestock owners to abandon water buffalo production. Rural syndicates along the Amazon in Brazil are not as well organized, so water buffalo herds continue to proliferate at the expense of crop farming.

How the drama of increasing water buffalo and cattle herds plays out over the next decade or so will also have major implications for maintaining biodiversity, particularly in the remaining floodplain forests. Ironically, it is the small farmers who adopt cattle ranching that are the most likely to trigger the greatest deforestation. Between Óbidos and Monte Alegre, the largest continuous stretches of floodplain forest are on large ranches. Many of the larger ranches in the vicinity of Santarém have passed through their major spasm of deforestation, and at least some of the remaining tracts of forest are likely to be left standing indefinitely.

Forest stands on large ranches are particularly valuable from the point of view of biodiversity for two reasons. First, the forest is in sizable blocks, which is important for sustaining populations of plants and animals that cannot easily disperse to isolated patches of forest more typical of areas dominated by small farms. The smaller the area of forest, the fewer species it is likely to contain. Second, only on these ranches are appreciable areas of forest still standing on the higher banks. On small farms, the banks of the Amazon and its side channels are usually cleared because they are prime cropland. If forest remains on small farms, it tends to be in backswamps. The species composition of forests changes from higher to lower elevations on the floodplain. Ranches thus contain some of the last remaining representative forest on high banks, at least in the middle stretch of the Amazon.

Smaller farmers are clearing forest on their land to accommodate their growing herds. Along the middle Amazon, the prospects for saving what is left of forests are brighter on the larger ranches—and policymakers need to include large landholders in any conservation plans. The incentive structure for maintaining forest on large ranches needs to be better understood, however. Some ranchers have expressed a desire to safeguard nature, and while their sentiments may be genuine, the current low price for beef may have more to do with the apparent slowing of deforestation on their properties. The degree to which communities or individual farmers might cooperate to safeguard sizable blocks of forest is uncertain.

The way in which small operators on the floodplain manage their cattle could provide insights on how to resolve the conflict between ranching and raising crops. Some farmers abandon crop farming altogether and dedicate their time to raising livestock, fishing, and taking on odd jobs in urban areas. As long as they take measures to protect their neighbors' crops, this approach makes sense. An example of this approach is a long-time farmer who now looks after the property of an absentee landlord of a small ranch near Santarém. As caretaker (*capataz*) he keeps twenty-seven head of his own cattle and thirty-six head for the landlord—and has decided it is no longer worthwhile growing crops. Instead, he now derives most of his income from fishing and ranching. Another small farmer with ten cattle grows most of his crops on a recently formed island near Santarém, not easily accessible to cattle or water buffalo. He tolerates damage by his livestock to a small plot of maize close to home, since the maize is mostly for domestic consumption rather than for market.

Cattle and water buffalo are here to stay on the Amazon floodplain, at least in the short term. Faced with that reality, improving the productivity of livestock and existing pastures might help alleviate pressure on remaining forests. Fiscal incentives might be made available to help farmers and ranchers intensify their cattle operations by improving pastures and installing sturdy fences. Some 11 million hectares of floodplain pasture are found along the Amazon, and some of them could be improved using low-cost techniques pioneered by small farmers and innovative ranchers, such as sowing or transplanting grasses preferred by large livestock.

Environmentalists might oppose such a suggestion, contending that it would only encourage a "sea of horns" on floodplains and uplands of the Amazon. But intensification is always preferable to extensification, the cutting down of more forest. Without subsidized credit or tax write-offs, few operators are likely to intensify production on a large scale given the current low prices for liveweight cattle. On Marajó Island, for example, farmgate prices for cattle were only sixty cents per kilogram in July 1996 and September 1997, a 50 percent drop from prices prevailing three years earlier. In the early 1990s, liveweight prices were fairly buoyant, and upland ranchers in many areas were investing in pasture improvement, among other measures, to upgrade their operations. Yet little attention has been paid to improving the efficiency of cattle operations on the floodplain, even during the boom years.

Forest-Friendly Small Livestock

While room exists for improving the productivity and management of cattle and water buffalo on the floodplain, a parallel effort should be directed toward finding viable alternatives for small and large operators alike. Alternatives would include fostering the adoption of small livestock that require little if any deforestation, and the domestication of native animals for meat production. Subsequent chapters of this book explore possibilities for developing more profitable, and environmentally benign, crop production systems.

A diverse array of small livestock is maintained on the Amazon floodplain, particularly by small producers. For the most part, small livestock inflict minimal environmental damage because extensive deforestation is not needed to support them, or because their numbers pale when compared to the population of cattle and water buffalo. In addition to large animals, some ranchers raise small flocks of sheep or goats. Amazon goats exhibit a diversity of colors and horn shapes (figure 4.9), rather than a single uniform breed. When they first arrived in the region is not clear. Goats are rarely mentioned in early travel accounts, although William Smyth and Frederick Lowe observed them in the 1830s at Gurupá on the lower Amazon.[7] Goats were probably not common in the Amazon until drought victims started arriving from the Northeast of Brazil in the late nineteenth century.

At low water, small herds of goats browse on shrubs, herbaceous plants, and grasses on floodplain meadows. In some parts of the world, such as the drier parts of Africa and the Middle East, goats have a reputation for destroying vegetation. But the scrubby trees of those marginal environments can be quickly stripped of their leaves. On the Amazon floodplain—at least in the numbers currently prevailing—goats do not appear to be wreaking any significant ecological damage. Similarly, the raising of pigs, chickens, ducks, and turkeys is largely compatible with the maintenance of forest cover. The only clearing typically performed for these livestock is to grow maize, their principal supplemental feed.

Sheep were extremely rare in the Brazilian Amazon in the early 1800s. Settlers from the arid interior of Ceará are largely responsible for introducing the animals to the region during the rubber boom. Sheep and goats are raised primarily for meat, rather than for wool or milk. Because sheep prefer to crop short grass, they are almost always raised alongside cattle on the Amazon floodplain. The flocks

Figure 4.9. A goatherd grazing on a floodplain ranch. Goats along the Amazon are derived mostly from stock introduced from the Northeast of Brazil. The diversity of coat color and horn shape indicates that a number of breeds are involved in the formation of herds. Fazenda Pensamento, Ilha de São Miguel, near Paracari, Pará, September 1993.

are generally small, and only a fraction of ranchers raise them. Both goat and sheep numbers appear to be increasing, however, and they may eventually exert significant pressure on grazing and browsing resources if their numbers become large.

Pigs, turkey, the helmeted guineafowl, and chickens are typically kept in pens or coops at high water. Ducks, of course, are able to fend for themselves during the flood, and pigs will feed on aquatic plants in shallow water. At low water, all small livestock are turned loose—except for pigs, which may remain confined to protect crops. Leafy vegetables, tomatoes, and spices for household consumption are generally grown on raised platforms, such as an old canoe, or may be encircled by a wooden stockade or an old fish net to keep perennially hungry fowl at bay.

Usually only cattle, water buffalo, sheep, and goats are taken off the floodplain at high water. During the Amazon's flood surge, sheep wander freely in villages, such as at Arapixuna and Vila Socorro, where they steer clear of backyards, which tend to be populated by vigilant dogs, and concentrate instead on soccer fields and the margins of trails. Villagers welcome the sheep both because they do not interfere with crops and because they help maintain the quality of soccer fields, such an important part of rural life on Sunday afternoons.

Small livestock make a welcome contribution to the table and income of floodplain farmers. Smaller livestock are able to fend for themselves much of the time and are also easier to market, since they can be taken to town on any of the passenger boats that ply the Amazon and its numerous side channels. Small farmers at Afuá on the northern coast of Marajó Island find a ready market for their free-range pigs in nearby Macapá. The swine are sold to a middleman who transports them in the hold of a sailboat to Macapá; there they are taken by pickup truck to butchers in town. One shipment from Afuá I witnessed in December 1994 contained forty-two pigs of diverse hues. On the island of Combu near Belém, Anthony Anderson and Edviges Ioris have found that pigs account for 10 percent of household income for smallholders during November, December, and January when açaí palms are no longer producing fruits in abundance.[8]

Traditional pig production on the floodplain is a far cry from industrialized operations in North America and Europe, where hogs are kept inside at all times and given commercial feed. The modernization of pig farming in developed countries has led to a rapid streamlining of breeds. Many hardy breeds that were adapted to foraging for acorns in woods and could be kept outdoors for much of the year are now rare or have become extinct. The large white pigs favored by commercial farmers cannot even be kept outdoors for long because they will suffer from sunburn. Pigs on the Amazon floodplain, in contrast, tolerate the sun well because they display a wide range of skin and hair colors, ranging from the Iberian black to the rusty-coated Alentejana from Portugal. Many are crossbred, and the lineage of some can be traced to the copper-colored Duroc, the red-coated Sorocaba, and black and white breeds such as the Saddleback or the Hampshire. Farmers on the floodplain are interested in efficiency and keeping costs down, rather than on maximizing yields.

It is precisely in the economic "backwaters" of the world that most of the genetic diversity of livestock and crops is found. The problem is that such areas are also among the poorest on earth. On the Amazon floodplain, poverty is probably not an appropriate description, but certainly the material standards of living are low. Most floodplain residents naturally wish to acquire more material goods and to have access to better educational and employment opportunities for their children. Eventually, much of the genetic diversity of pigs, as well as other livestock, is likely to be lost as the region continues to develop. A major challenge facing development planners, then, is how to maintain and even increase the biodiversity of cultural landscapes while helping people improve their living standards.

Forest and Meadows as Livestock Feed Stores

A large number of floodplain plants contribute indirectly to human nutrition by providing food for cattle, pigs, chickens, ducks, and turkeys. Virtually all floodplain residents keep some livestock, and securing feed and fodder for domestic animals is a major preoccupation.

Small livestock are fed table scraps and maize, either purchased or home grown. But a significant portion of their feed is gathered in the wild, particularly in the floodplain forest. Several fruits are gathered from floodplain forest to feed small livestock. The oily fruits of the urucuri palm, for example, are fed to pigs, chickens, Muscovy ducks, and turkeys (figure 4.10). Urucuri palms, also known as arucuri in Amapá, bacuri in Mato Grosso, and cocinho in Rondônia, are often left standing in fields and next to homes because of their useful fruits. For the most part, the fruit, which appears in abundance between November and May, is fed to livestock on the floodplain, but some consignments (*encomendas*) are sent to urban dwellers to feed their backyard pigs, as I observed in Macapá in December 1994. Urucuri is ubiquitous on the middle Amazon floodplain, where its broad,

Figure 4.10. Fruits of urucuri palm gathered in a home garden for pig feed. The urucuri palm was left standing near the home because of its useful fruits. Arapixuna, Pará, May 1996.

arching fronds attain almost six meters and lend grace and beauty to the landscape. Urucuri also occurs in widely scattered upland locations, ranging from Amapá and Pará to Acre and Rondônia in the west. It prefers damp sites with relatively fertile soils in upland areas, but does not appear to be as important to the household economy there as on the floodplain.

The fruits of other floodplain palms are also gathered to feed small livestock, especially pigs. Jauari, for example, an aquatic relative of tucumã palm, conveniently fruits at high water from February to August, when pigs are confined mostly to raised wooden corrals. It often occurs in stands in various water types ranging from the muddy Amazon to clear (Tocantins and Xingu rivers) and black water (Rio Negro). Along the Igarapé Jari, which connects the Amazon to the lower Tapajós, fathers and sons steer canoes up to partially submerged jauari palms and dislodge the green, golf-ball sized fruits with their paddles or poles. Care is taken not to brush against the trunk or fronds of the palms because they are armed with needle-sharp spines. The fruits float when they fall into the water and can be easily scooped up and loaded into the canoe. Along the Paraná Nhamundá near Terra Santa, farmers stockpile jauari fruits in open containers or boxes to feed livestock and for fish bait.

The knobby fruits of bussú palm are fed to pigs in the estuarine area, such as around Afuá on Marajó, even though they contain meager amounts of pulp. The main use of bussú is for thatch, but the hard core of the fruit can be chopped open with a machete to obtain emergency drinking water. Açaí fruits, usually destined for urban markets or domestic consumption, are occasionally fed to pigs, such as in Murumuru near Santarém, especially if transportation cannot be arranged to take the fruits to market.

When the tide is out in the estuarine area, farmers generally allow their pigs to wander through the forest to feed on fallen fruits. On Combu Island, for example, pigs feed voraciously on the juicy fruits of yellow mombim and buriti palm. Buriti fruits are also an important food for certain game animals, especially gray brocket deer, red brocket deer, and tapir. Along the middle and lower Amazon, however, such animals are now rare or locally extinct due to overhunting and habitat destruction; they have been largely replaced by pigs that trace their ancestry to the rainforests of Southeast Asia.

American oil palm occurs sporadically along the Amazon from roughly the border between Amazonas and Pará to the Peruvian Amazon. Caiaué, as the prostrate palm is known in the Brazilian Amazon, is occasionally harvested for its oily fruits to feed to pigs or to prepare folk remedies. Introduced to the Amazon from Central America in precontact times, the American oil palm has evolved various uses as different cultures and animals have come to the region.

Several other sizable trees of the floodplain forest produce fruits or nuts that are fed to small livestock. The reddish-brown pulp of the fruit of the cannonball tree is fed to pigs and poultry. Known as *castanha de macaco* along the middle Amazon and as *ayahúman* in Peru, this relative of the Brazil nut tree bears its fruits on the trunk and main branches, rather than at the end of branches. The trunk soon divides into a number of gently curving branches, making it relatively easy for teenagers to climb and twist off the pendulous fruits. The nuts are encased in a hard, round shell that must be split open with a machete or ax to access the flesh. The fruits, which are harvested as the waters rise from March to June, are too heavy to catch; if the floodwaters have already entered the forest, the medicine-ball–size fruits are dropped into the water where they will bob until picked up (figure 4.11). Nuts of the related sapucaia tree are also chopped up and fed to chickens, especially to delicate chicks to help fortify them. Because of their usefulness for livestock feed, sapucaia and cannonball trees are often spared when people clear home sites along the banks of the Amazon.

Residents on the floodplain are often forced to pen their chickens at the height of the flood and fruits gathered in the flooded forest provide valuable nutrients for the confined birds. Uruá, another denizen of floodplain forests, is normally employed as fish bait, but is fed to poultry on farms along the middle Amazon, such as along the Igarapé Jari near Arapixuna and at Amador, about mid-way between Santarém and Óbidos.

As the waters recede, numerous plants arise from buried or wind-blown seeds or tubers to blossom and reproduce before the next flood, and some of these provide poultry food. Apé, for example, produces small, round tubers that are dug up to feed ducks (figure 4.12). In the past, hunters and gatherers may have collected the tubers at low water. Early farmers on the floodplain conceivably cultivated the plant alongside other root crops such as sweet potato, manioc, and the New World yam. Apé may be one of those plants that has slipped in and out of domestication with the passage of cultures.

Figure 4.11. Fruits of the cannonball tree gathered in floodplain forest to feed pigs and chickens. The boy and his father work as a team to gather the fruits, which emanate from the trunk, rather than the tips of branches. Carariacá, Pará, May 1996.

Figure 4.12. Tubers of apé, which grows wild in open habitats on the floodplain, are fed to ducks. These tubers were gathered in the boy's home garden, which has recently emerged from the annual flood. Igarapé Jari near Arapixuna, Pará, August 1993.

Patazana, both a weed and a resource, proliferates in open spots. Farmers hack the herbaceous plant down to make room for crops, but they also harvest some of the stems and leaves to feed cattle. The tall, spindly plant is a welcome guest around houses where women and children strip off the peppercorn-sized seeds, which resemble Job's tears in the grass family, to feed chickens (figure 4.13).

Potential for Domesticating Game Animals

Along the Amazon, most livestock sounds that greet the dawn are made by Old World animals—the melancholy lowing of cattle, the impatient snorts of horses, and the incessant baying of donkeys, the latter used mostly to produce mules. The bleating of sheep and goats, though not common sounds, are also echoes from the Old World. The startled, cackling call is of helmeted guineafowl, known in Brazil as *galinha d'Angola* (the chicken from Angola), which were brought from sub-Saharan Africa to serve as sentinels as much as food. The fussing of chickens and the deep grunts of pigs, so characteristic of village life in the Amazon, also reverberate on farms and in the jungles of Southeast Asia where they were domesticated.

Only the frantic gobbles of the turkey, the emphysemic wheezing of the Muscovy duck, and the excited, high-pitched whistle of the black-bellied tree duck (figure 4.14) are New World sounds. And of them, only the two ducks can be

Figure 4.13. Seeds of patazana, growing spontaneously in a home garden on the floodplain, are fed to chickens. The young woman works on her parents' farm; she recently tried her hand as a teacher in bustling São Paulo, but has decided that country life in her native land is less agitated. She represents an unusual countercurrent to the more common trend of rural-urban migration. Igarapé Jari near Arapixuna, Pará, August 1993.

Figure 4.14. A captive black-bellied tree duck on the Amazon floodplain. These tree ducks are sometimes raised in backyards along the Amazon. Paraná Cachoeri near Oriximiná, Pará, June 1994.

considered Amazonian. The time is ripe to explore the potential of other native animals for meat and leather.

I do not wish to suggest that people should abandon Old World livestock en masse to adopt game farming. Rather, the purpose of this section is to explore the potential for domesticating some native animals for "niche" markets. The idea is to increase options for farmers, not to suggest that cattle, sheep, goats, pigs, and chickens are inappropriate. Native animals, on the other hand, are likely to be less destructive to the environment than livestock production based on cattle and water buffalo.

The meat of the capybara, which resembles a giant aquatic guinea pig, is savored along the Amazon and other rivers in northern South America (figure 4.15). Long ago, Catholic priests declared that capybara was not a mammal, thereby conveniently overcoming the Catholic stricture against eating red meat during the months leading up to Easter. And today in the llanos of Venezuela, some ranchers harvest wild capybara, particularly during Lent.

A prolific breeder, the aquatic rodent reaches a market weight of thirty-five kilograms within about eighteen months and produces four times as much meat as cattle on a per hectare basis. Along the middle Amazon, capybaras are shot indiscriminately and are no longer common; they are also heavily hunted along tributaries of the Amazon, such as the Tocantins. Although technically illegal, because the commercialization of wildlife products is prohibited in Brazil, capybara meat is occasionally sold "under the counter" in such markets as Manacapuru and Itacoatiara.

At least one farmer on an island near Santarém has taken the initiative to domesticate capybara, keeping a small breeding herd since the 1970s. By 1993, the farmer had sixteen capybaras of various ages maintained in several pens. The capybara are fed wild grasses, including pemembeca, wild rice, and murí. Maize is the main supplemental feed and capybara droppings are placed on vegetable plots tended by the farmer's wife. The German shepherd-sized rodents reproduce twice a year and deliver one to eight offspring each time. Capybara herds build up quickly because captive capybaras reach sexual maturity in fourteen months.

Other than concern about being raided by the wildlife protection service (IBAMA; Instituto Brasileiro do Meio Ambiente e dos Recursos Naturais Renováveis), the only misgivings the farmer expressed about raising capybara is that they are so affectionate, they become like pets. And that poses a dilemma when the time comes to pay back the farmer for all his trouble gathering wild grasses.

It is ironic that this innovative farmer must breed his animals in a clandestine manner. He cannot legally sell capybara meat, even to his neighbors. His operation does not involve deforestation, yet nearby ranchers can clear as much forest on their land as they see fit in order to raise introduced cattle and water buffalo. Regulations in Brazil that prohibit the commercialization of wildlife should be amended to allow for the sale of animals that have been raised in captivity. By lifting the ban on the sale of capybara meat, farmers and ranchers would have an incentive to domesticate the prolific creature, or at least manage wild populations. The wildlife service is largely ineffective at controlling the slaughter of wild ani-

Figure 4.15. A young capybara, kept as a pet on farm on the Amazon floodplain, wears a conspicuous plastic collar to alert people that it is a pet; otherwise it would likely be shot if it strayed from home. At night, the animal, which will likely triple in size, has to be tied under the house because it tries to climb into bed with the children. Surubim-Açu, near Santarém, Pará, September 1993.

mals and the destruction of their habitats, anyway, so why not use market forces to encourage the domestication of at least some of the overexploited game species?

Another rodent native to the floodplain, but largely unknown in urban areas is the Amazon bamboo rat. Known as *saiuá* along the central Amazon, the plump rat is both a pest and a delicacy. It is fond of climbing maize to nibble on the cobs before they are ripe. A recently arrived sharecropper from Paraná who lives near Alenquer complained that the nocturnal rat was causing appreciable damage to his maturing maize. When I asked whether the rat was eaten, he replied in a somewhat disparaging tone that there are people who eat them (*"tem gente que come"*). I subsequently learned that long-time residents of the floodplain, such as at Piracauera da Cima upstream from Santarém, set wooden traps for the secretive creature, baited with maize or sweet manioc. The quarry is then eaten. The Amazon bamboo rat might one day be raised in and around homes, in a manner similar to guinea pigs in the Andes.

Giant river turtles (*Podocnemis expansa*), once penned in the thousands by indigenous groups along the Amazon and some of its tributaries, have never been domesticated on a commercial scale. Attempts to raise turtles in captivity have proved largely futile, at least from the entrepreneurial perspective. In the 1970s, a rancher received fiscal incentives from the regional development agency (SUDAM) to raise the giant river turtle on his property, Fazenda Lago Pretinho, a few kilometers upstream from Juriti. The federal wildlife protection service, IBAMA (then called IBDF; Instituto Brasileiro de Desenvolvimento Florestal), arranged for 70,000 turtlings that had been taken from one of the region's protected beaches to be released into the ranch's dark-water lake. The murky lake, which empties onto the floodplain through a narrow gap, was at one time staked to prevent the turtles from escaping. On my visit to the ranch in 1994, no one knew how many, if any, turtles were left in the lake. A modest investment in research to better understand the natural history of river turtles, and requirements and constraints under captive conditions, might eventually provide a handsome payoff. Local fishermen and farmers can readily provide a list of fruits favored by the giant river turtle, which approaches the size of a hula hoop and sells for well over one hundred dollars in clandestine markets.

5

Islands in the Sun

It is ironic that one of the most fertile areas in the Amazon for crop production is largely unutilized for that purpose. The area of the Amazon floodplain in crops is a fraction of that devoted to pasture for cattle and water buffalo. The growing cities in the Amazon are being fed largely from farms in other parts of Brazil and abroad. Rice, an important staple consumed in Amazonia, is produced mostly in upland fields of central Brazil. And although bread is a significant source of protein for city folk in Amazonia (partly as a result of government subsidies), much of the bread consumed in the Amazon does not benefit wheat growers southern Brazil; instead, much of the wheat consumed in the country is imported.

The Amazon floodplain is nevertheless a favorable environment for intensive food production. The annual floods deposit a fresh layer of alluvium that rejuvenates the soil, which allows farmers to cultivate the same plot every year without resorting to fertilizers to shore up yields. Farmers can also take advantage of nearby water for irrigation. Although many crops, such as manioc and sweet potato, produce well without watering, yields of maize and vegetables increase when the soil is kept moist during the dry season. Finally, the availability of relatively cheap water transportation greatly facilitates the marketing of crops. This chapter focuses on practices, problems, and prospects involving annual crops.

Two myths about floodplain farming need to be put to rest. The first is that the annual floods are a devastating phenomenon that undercut any hope for intensifying agriculture. All farming environments have their challenges; the question is how to adapt practices to overcome natural hazards. Farmers and livestock owners in Amazonia employ many adaptive strategies to roll with the punches rather than fight them. Second, floods do not completely obliterate weeds, allowing farmers to start each season with a clean slate. Although it is hard to generalize because of the vast array of different cropping systems, soil types, and surrounding plant communities, floodplain farmers are probably no better off with regard to weeds than their upland counterparts. The very fertility of many floodplain soils promotes rapid weed growth. It is true that some bars, spits, and low-lying islands emerge from the floods with few plants, but they are rapidly colonized by seeds deposited by wind, birds, fruit-eating bats, or the receding waters. On higher banks, which dry out first as the flood wanes, some weeds are already growing in the shallow water before the farmer has a chance to plant crops. Murí grass, for example, grows quickly on high banks and usually needs to be hoed and burned to make room for crops. Farmers sometimes weed manioc as they harvest in order to have a cleaner planting surface once the flood recedes.

The burning of grasses and other weeds on the banks of the Amazon and along lake margins serves several purposes. Space is provided for crops, ash from the fires quickly releases some of the nutrients bound in the weeds, and the cleared area provides nesting sites for the yellow-spotted Amazon turtle. The oblong-shaped eggs of *tracajá*, as the species is known locally, are relished in town and countryside. Some attribute aphrodisiac properties to the eggs, found forty or so to the nest. A farmer near Urucurituba on the floodplain a little upstream from Santarém volunteered that he was being particularly fastidious with weeding along the edge of his manioc field on the bank of the Amazon to "call" (*chamar*) the wary turtle. Along the banks of the Maicá, a river draining an Amazon floodplain lake that abuts the uplands some twenty kilometers downstream from Santarém, farmers burn vegetation to encourage nesting by this turtle and pitiú, a smaller, related species. Small patches of riparian vegetation are burned to encourage turtle nesting even in areas not planned for crops.

Both the yellow-spotted Amazon turtle and pitiú are sold clandestinely in urban markets along the Amazon for use in a variety of dishes. In Santarém's main market, for example, three pitiú turtles, hidden in a sack inside a Styrofoam container, were for sale in July 1996. It is common knowledge in the fish section of the market who has turtles for sale on any given day.

In upland areas, farmers clear forest or second growth after yields decline in their swidden fields because of weed invasion and exhaustion of soil nutrients. Farmers typically set fire to their fields a month or so before the rainy season. In this manner, they avoid being caught with piles of plant debris in their fields before it has a chance to burn. The ash provides a jump start for weeds and crops alike. Farmers also practice slash-and-burn farming on the floodplain, although cutting and burning is synchronized with the flood cycle rather than with the wet and dry seasons.

For the most part, crop production is not a significant force in deforesting the floodplains. A small plot of land may occasionally be cleared for crops such as maize or manioc, but farming is largely carried out on already-cleared land. During the colonial period, some floodplain forest was cleared for cacao, mainly along the middle and lower Amazon and during the heyday of jute, from roughly the mid-1930s to 1980, appreciable sections of floodplain forest were cleared in some areas of the middle Amazon. But according to one farmer at Arapixuna near Santarém, senile cacao orchards, rather than forest, made way for the jute boom.

Before we explore how farmers husband their crops in response to the rhythm of the floods, a brief overview of cropping patterns is in order. Three main categories of farmers are found along the Amazon: those who till only the floodplain; those who cultivate both the upland and floodplain; and those who plant only on the uplands. The focus here is on the first two categories.

Living arrangements of farmers who cultivate only on the floodplain vary widely. Some spend the entire year on the *várzea*, especially the relatively privileged families living on high banks of the river's main channel. Such locations may be inundated only for a few weeks, or only during exceptionally high floods. Other year-round residents live along the margins of lakes in the interior of floodplain islands, and tend to be poorer because transportation is more sporadic and the growing season is much shorter. Farmers on less-privileged parts of the floodplain tend to focus on crop varieties, particularly of manioc, that have been honed for generations to grow rapidly during the low-water season. Some farmers abandon their homes at high water and camp out on uplands, such as on the savannas inland from Arapixuna. They and their families set up makeshift huts, usually with fronds of the ubiquitous curuá palm, and seek off-farm employment or dedicate themselves to fishing. The high-water season for them is a time of deprivation and suffering because they must purchase most of their food.

Farmers who cultivate both uplands and the floodplain are in a more privileged position. They literally have the best of both worlds, and are emulating the practices of many indigenous cultures before them. At low water they plant a mix of upland and floodplain varieties of manioc, the basic staple, on banks of the river and side-channels. Families can harvest crops from *terra firme* fields while the Amazon is on its annual rampage. Their main residences are typically in villages on upland bluffs overlooking the Amazon, and the upland fields may be some distance from home, often referred to as in the "colony" (*na colonia*) or in the "center" (*no centro*). At low water, the families may lock up their houses and shift to simple huts on the floodplain to tend crops until the waters rise again.

Living arrangements, however, are fluid. During exceptional floods, many residents who normally spend the entire year on the *várzea* become temporary refugees. They seek out relatives in towns and cities and may spend several months as guests until waters have subsided to at least the floorboards in their houses. It is not uncommon for floodplain farmers to have relatives in urban areas; their older children are often already living much of the time in urban areas because of schooling. Burgeoning shantytowns on the perimeter of towns and cities along the Amazon become even more crowded when the river-sea surges higher than usual.

Whether farmers live year-round on the floodplain or part of the time on uplands, fields are sometimes located appreciable distances from home. Pressure from cattle or water buffalo may force a farmer to plant crops on a new island, far from livestock. Or, arrangements may be made to till part of someone else's property, where cattle and water buffalo are less likely to cause damage. Some families plant annual crops near their home, but also take advantage of opportunities created by newly formed islands, bars, and spits further afield. Farmers who live in upland villages on the floodplain margin may travel as far as several hours by motorboat to reach their seasonally occupied property.

Only a few of the annual crops grown on the Amazon floodplain will be highlighted here. The profiled crops were chosen to illustrate several basic principles concerning agricultural development. First, reliance on a single crop as the main source of income is a risky proposition. Jute provides a valuable lesson in the dangers inherent in excessive dependence on a single cash crop. Second, as agriculture becomes more intensive and market-oriented, the need for high-quality planting stock increases. Commercial varieties often require high levels of purchased chemicals, especially pesticides, to outyield traditional varieties. The experience of vegetable growers illustrates the mismatch between varieties available to them and local ecological realities. Third, agricultural intensification should be tried out on a limited area first, in order to iron out at least some of the major problems before attempting to promote "modern" practices on a wide scale. This can been seen in the different approaches to irrigated and mechanized maize and rice production on the Amazon floodplain. Finally, farmers should deploy a diverse array of varieties of their traditional food crops, for example, manioc and squashes, to avoid risk.

The Rise and Fall of Jute

Jute was introduced to the Amazon in 1931 and soon became the dominant cash crop along the middle stretch of the Amazon floodplain. A Japanese entrepreneur, Isukasa Uetsuka, pioneered jute cultivation by bringing in some Japanese settlers to grow the crop and by founding the first jute plant, Companhia Industrial Amazonense, near Parintins. Jute cultivation spread relatively slowly up the Amazon from Parintins, reaching Peru by the 1950s. By the 1970s, the Shipibo, an indigenous group living along the middle Ucayali, had adopted it as a cash crop. At the height of the jute boom in the mid-1970s, the crop eventually occupied approximately 60,000 hectares of floodplain soils along the Amazon, mostly between Manacapuru and Monte Alegre. Much of the jute was destined to make sacks for coffee and sugar.

By 1971, an upland fiber crop, urena, known locally as *malva*, was also cultivated on higher parts of the Amazon floodplain in Amazonas and Pará. Urena is an Old World plant that originated either in Africa or India. When and how it was introduced to Brazil is uncertain, but it has become naturalized in many tropical parts of the country. Urena lingered as a weed in eastern Pará until jute took off along the Amazon floodplain. Jute effectively pulled along the domestication of feral urena. During the heyday of natural fibers for coffee sacks, most of the urena in Brazil came from the upland fields of eastern Pará, but modest amounts were also grown on the Amazon floodplain about a decade before jute started its decline.

By 1953, Brazil no longer had to import jute from India and Pakistan, and by 1960, Brazil emerged as the world's third largest jute producer, all of which was grown on the Amazon floodplain. By the early 1980s, however, competition from synthetic fibers, particularly polypropylene, had undermined jute's market. In Brazil today, synthetic bags are one-third to one-half the price of those made from jute, and are widely used for many commodities, particularly for groundnuts, potatoes, and onions. A shift to bulk handling of many commodities, particularly sugar, grains, and soybeans, has accelerated jute's collapse. Finally, producers in Bangladesh have driven down the price of jute fiber because of their relatively low operating costs; Bangladesh now provides nearly half of all jute traded on world markets.

The crash of jute as the mainstay of crop farmers along the Amazon has provoked a scramble for alternative livelihoods, with many former jute farmers turning to cattle raising and fishing. Farmers cleared appreciable tracts of floodplain forest and cacao orchards to grow jute over the crop's fifty-year boom; more forest will succumb if farmers orient more of their operations to cattle and water buffalo.

Miguel Pinto, a seventy-year-old farmer at Arapixuna near Santarém, recalls the changing landscape along Igarapé Jari in which jute has played a part. The sinuous, creamed-coffee–colored side channel of the Amazon flows in front of Miguel's village and connects the turbid Amazon with the blue-water Tapajós. When he was a youngster, Miguel remembers extensive cacao orchards along the banks of the Jari, but jute farmers and the great flood of 1953 did away with the groves. Today, it is doubtful whether a single cacao tree remains there, and jute is also gone. The landscape is currently dominated by meadows with grazing cattle and water buffalo interspersed with overgrown home gardens marking the sites of abandoned farms; stands of lime-green Amazon willow on mud flats; a few patches of floodplain forest that have been logged over; and scattered small farms with a patchwork of bananas, beans, maize, and manioc.

A similar story is told in many other parts of the floodplain. Some 10 kilometers downstream from Óbidos, for example, a farmer recalls his home being surrounded by an extensive cacao grove when he was a boy. The farm, on the north bank of the Amazon near the entrance to Paraná Mamauru, one of the river's side arms, had been occupied for generations, and the cacao grove was kept productive by replacing older trees with newer ones. With indifferent prices for cacao and a growing market for jute, however, much of the cacao was cut down in the 1940s. Groves that had survived the 1953 flood were largely replaced with jute by the early 1960s. Now, the landscape around the farm is relatively open, dominated by meadows for thirty-five cattle. Their other major source of income is fishing, to supply the ready market in Óbidos. Jute is no longer grown on the farm or anywhere in its vicinity.

Shuttered jute plants that used to press and bale the fiber for export to other parts of the country are visual evidence of the crop's demise. Manaus once had seven jute plants, but only two were still operating by 1991. All three jute factories in Santarém, located along the waterfront in the eastern side of town, have shut down. Thousands of shareholders lost their investments when CATA (Companhia Amazônia Textil de Aniagem), FIBRASA (Fibra da Amazônia, S.A.), and TECEJUTA (Companhia de Fiação e Tecelagem de Juta de Santarém) were boarded up. TECEJUTA, the largest jute factory in Santarém, was the last to close its doors, in 1986. In 1970, Óbidos had five plants, but the last one went out of business in 1992 (figure 5.1). Parintins, where jute cultivation started, no longer had any jute factories by the early 1990s. Itacoatiara once had four companies purchasing and pressing jute in the early 1980s, but had none in 1991, and at least one of the former factories has been converted to a warehouse. Jute has followed the familiar boom-and-bust cycle of so many economic activities in Amazonia.

Figure 5.1. A closed jute factory in Óbidos, Pará, in June 1994. The factory opened in 1941 as the Companhia Paulista de Aniagem, and changed its name to Usina Amazônia when purchased by new owners in 1964.

The demise of jute has also exacerbated urban poverty. The shuttering of jute factories has contributed to urban unemployment and the abandonment of numerous farms on the floodplain. Some of the farms were sold in haste, and for relatively low prices, to ranchers. On Ilha Grande, downstream from Óbidos, a rancher remarked that one can easily spot where a jute farmer used to live: a clump of ancient mango trees on the river bank where the home used to be, now set in a sea of pasture. Rather than try their hand at cultivating the *terra firme*, most farmers leaving the floodplain apparently head for urban areas where they typically settle in sprawling slums. So many people are crowding in to Alenquer, with approximately 30,000 inhabitants, that the town's only airstrip had to be declared off limits as huts encroached ever closer to the runway. Pilots intending to fly to Alenquer now file flight plans for nearby Monte Alegre or Óbidos. Although the airport is officially closed, small planes occasionally land on the narrow strip after several low passes to warn away pedestrians taking shortcuts, children playing, and grazing livestock. The urban fringe of virtually all towns and cities along the Amazon is growing, mostly with squatter settlements. Locals refer to the shantytown on the eastern part of Óbidos, officially known as the "new city" (Cidade Nova), as the suburb of jute refugees (figure 5.2).

Some government officials in Santarém and Monte Alegre talk enthusiastically about the imminent return of jute as a commercial crop. The desire of well-to-do consumers for "natural products" is often cited as reason for this optimism. Proponents of jute in the Brazilian Amazon tout the crop as "ecologically correct" because polypropylene, jute's main rival, cannot be easily recycled. Jute fiber can be incorporated in some clothing, but it is doubtful whether such scratchy garments can capture a sizable market; wearing clothes made from jute is akin to donning the sackcloth of repentance. A few farmers, as on Carmo Island down-

Figure 5.2. Dom Floriano Street in Cidade Nova, a shantytown on the outskirts of Óbidos, Pará. A makeshift wooden bridge separates "New City," where many former jute farmers live, from the older part of town. June 1994.

stream from Óbidos, still cultivate small patches of jute because the coarse fiber is useful for caulking canoes.

Yet jute's time along the Amazon has essentially come and gone. In 1991, the Brazilian government eliminated import tariffs for the fiber. Several former jute farmers I talked with do not long wistfully for the return of the Asian crop. They remember arduous days cutting and stacking jute in knee-deep, murky water, the lair of stingrays, some the size of dinner tables. Other hazards include severe jolts from electric eels; piranhas, however, are not a concern, several B-grade movies to the contrary. After soaking for a week or so, the loosened fibers had to be torn off the stems by hand. One newspaper in Belém described jute cultivation as a "tuberculosis factory" on account of working long hours in damp conditions.[1]

While no single crop is likely to take the place of jute, a mix of crops is helping fill some of the economic vacuum left by jute's retirement from the agricultural scene. Overall, though, average per capita incomes on the *várzea* have probably declined in the last decade or two. Indeed, some former jute land is reverting to second growth or is being converted to cattle pasture. The total cultivated area on the Amazon floodplain has therefore declined since the jute boom waned in the mid-1980s. That no single crop is replacing jute is probably a healthy sign. Farmers will be less susceptible to wild swings in commodity prices and to catastrophic outbreaks of diseases or pests.

The Emergence of Market Gardening

Vegetables have emerged as one of the most promising options for former jute growers near urban centers. Women, in particular, have specialized in vegetable production (figure 5.3). With the spectacular growth of cities, demand for a wide variety of vegetables, particularly tomatoes, lettuce, cabbage, cucumbers, bell peppers, okra, and spring onions, has increased dramatically. Traditionally, people in Amazonia have not eaten many vegetables, but customs are changing with the rise of the middle class and the influx of people from other regions. Salads are now much more commonly served in homes, restaurants, and fast-food outlets in Amazonian cities.

With the paving of the Belém-Brasília highway and the construction of the Transamazon highway in the early 1970s, growers in southern Brazil initially supplied most of the demand for vegetables in such towns as Belém, Santarém, and Manaus. Locally grown vegetables have come mostly from intensively managed farms on uplands on the outskirts of cities. Now the picture is changing, at least in the middle Amazon, where floodplain farmers appear to be increasing their market shares in urban areas. For example, they now supply virtually all of the tomatoes entering the markets of Santarém.

Commercial vegetable growers tend to cluster along certain stretches of the river. At Piracauera, an hour's boat-ride upstream from Santarém, for example, farmers concentrate on the Rasteiro variety of tomato, the Japonesa cultivar of watermelon, and honeydew melons, as well as leafy vegetables for the Santarém market. The community of a few dozen dwellings is relatively prosperous because it is on a high bank of the Amazon, is served by daily boat traffic, is close to a major market, and has focused on varieties in demand by the more affluent shoppers.

Figure 5.3. A farmer harvesting lettuce in a platform vegetable bed on the Amazon floodplain. Paraná Cachoeri near Oriximiná, Pará, 20 June 1994.

Figure 5.4. Lettuce and other vegetables growing in cured cattle manure on raised beds on the Amazon floodplain. Growers using elevated beds can grow vegetables year-round. Paraná Cachoeri near Oriximiná, Pará, June 1994.

Another hot spot for vegetable production is along the silt-laden Paraná Cachoeri, a side arm of the Amazon near Oriximiná. There farmers have specialized in growing lettuce, cabbage, spring onions, and coriander in beds raised on stilts (*canteiro suspenso*) (figure 5.4). Most of the vegetables are destined for markets in nearby Oriximiná and Porto Trombetas, about a four-hour boat-ride away. Mineração Rio Norte, the bauxite mining company based at the latter town, provided the main catalyst for vegetable production along Paraná Cachoeri two decades ago. Many of its employees are from the larger cities in the Amazon and temperate regions of Brazil, where salads and vegetables were a regular part of their diet. In order to reduce the frequency of costly charter flights laden with vegetables from southern Brazil, the company drew up contracts with growers along Paraná Cachoeri to produce vegetables for shipment to Porto Trombetas. In 1994, the largest producer, José de Oliveira Lima, had seventy-four elevated vegetable beds in production, some as long as fifty meters.

Vegetable farming on the floodplain has forged links with other land-use systems, especially cattle ranching and the timber industry. Cattle dung is incorporated in topsoil along with wood ash from kitchen stoves or manioc ovens, or is used "pure" after curing for a few months. The manure is either obtained free from local ranches or is purchased at the modest price of twenty-five cents per fifty-kilogram sack. Cattle are usually corralled at night to reduce theft and facilitate milking, and overnight droppings are easily bagged in polypropylene sacks. Buyers from Santarém range as far as Alenquer to buy manure from floodplain ranchers for sale to upland vegetable growers.

Cattle and water buffalo thus present a dilemma for vegetable growers. On the one hand, constant vigilance is required to make sure that large livestock do not trample fences; on the other, their manure is a good fertilizer. One grower near Urucurituba was using sawdust from Santarém to spread around his tomato plants. While the sawdust does not supply significant quantities of nutrients, it helps suppress weeds and conserves soil moisture. Sawdust is obtained free from the twenty-three sawmills operating in Santarém, a dramatic increase from twenty years ago. The proliferation of sawmills there and in many other Amazonian towns and cities is a response to the greater access to upland forests along pioneer roads and soaring prices for all types of hardwood for both the domestic and international markets.

Vegetable production is a highly intensive operation on the Amazon floodplain because it requires a great deal of labor and agrochemicals. Almost all the vegetables grown on the floodplain are produced by small farmers with heavy reliance on family labor (figure 5.5). Women and children help transplant, weed, and harvest these delicate crops. Commercial growers often employ workers, including women and children, at the rate of ten dollars a day, particularly for harvesting tomatoes.[2] Vegetables are grown in elevated beds as well as on the ground (figure 5.6).

Maçaranduba is one of the preferred woods for elevated beds, which are generally twenty-five to fifty meters long and a meter wide. Rot-resistant maçaranduba grows in upland forests and the floodplain. The beds are surrounded by raised plank walkways to facilitate bed preparation, watering, application of pesticides, and harvesting. The greater investment in such beds is compensated by the year-round production. Farmers surround the base of the wooden stilts with tin guards fashioned from cans to prevent rats from climbing up. As an additional precaution against rats, one grower along Paraná Cachoeri also deploys five cats, who have grown accustomed to not setting paw on dry land for several months each year. The almost universal black rat is probably the main culprit, since the platform beds are built close to home where the introduced rodents spend much of their time, although some indigenous rats may also be involved. Raised platforms are an adaptation from the ancient custom of growing herbs and spices in small, elevated beds by the kitchen.

When vegetables are grown on the ground, the cropping season is only about six months, at low water, and the plots must be fenced to keep out livestock. Some farmers, such as those on Ituqui Island, downstream from Santarém, opt for both elevated beds and vegetable plots on the ground. Because of their high value, vegetable growers often invest in irrigation systems and acquire stocks of pesticides. Many growers have purchased four- or five-horsepower diesel pumps to water their vegetable plots. In some cases, several families cooperate to buy and maintain portable irrigation pumps. Irrigation allows two tomato crops to be grown on the *várzea*.

Although the spread of vegetable farming along the floodplain is undoubtedly improving rural incomes and employment opportunities, as well as helping to improve the vitamin and fiber content of diets in urban areas, market gardening raises ecological issues. Because heat and humidity create ideal conditions for the proliferation of insects and pathogens, vegetable production in the tropics is renowned for the heavy use of agrochemicals, particularly pesticides. Insecticides and fungicides are often an attractive method to control damaging pests, at least in the short term. While insecticides have undoubtedly increased food produc-

Figure 5.5. Family tending a nursery on the floodplain for capsicum pepper, known locally as *pimenta de cheiro* used mostly to flavor fish dishes. The fish netting is designed to keep out chickens. Sítio Camanova, Piracauera de Cima, near Santarém, Pará, July 1996.

Figure 5.6. A floodplain vegetable farm employing elevated platform beds as well as spring onion plots on the ground. The property is fenced to keep out cattle. Piracauera de Cima, near Santarém, Pará, September 1993.

tion and improved farmer income, hidden costs are involved in their widespread application. Apart from possible poisoning of farmers and workers due to inappropriate handling of pesticides and poisoning of consumers due to insufficient washing of vegetables, the use of such chemicals in floodplains raises the specter of contamination of fish, the most important source of animal protein for the regional population. One effort to intensify production of annual crops in the vicinity of Praia Grande on Marajó Island led to pesticide contamination of water courses and allegedly impaired the health of some individuals.

The environmental impacts of pesticides can be mitigated, as long as the chemicals are used judiciously. Unfortunately, farmers along the Amazon receive little if any instruction on the proper handling and application of the poisons. Information is available from agricultural supply stores in the larger towns, but many growers are illiterate, so solutions may be prepared in the wrong concentrations and handled in an unsafe manner.

Tordon is the pesticide of choice for tomato growers in the vicinity of Santarém. But other pesticides are probably deployed, depending on availability and price. Vegetable growers in Roraima are using a veritable arsenal of highly toxic pesticides to protect their crops, including DDT, Baygon, Nevapol, Nirosin, Actiban, Permosi, and Malatol. DDT is banned in the United States, but the persistent pesticide is exported to a number of developing countries. In the Brazilian Amazon, it is still sprayed in the interior of buildings in an attempt to control vectors for malaria, a practice started at the close of the Second World War. The regulatory environment governing the use of pesticides appears to be lax, at least in the Amazon.

The issue of insecticides on the floodplain has not caught the attention of public health officials because vegetable growing is still a small-scale affair, and the enormous volume of water in the Amazon is likely diluting harmful chemicals. At the moment, high mercury levels in some fish, a result of widespread gold mining, is justifiably claiming the attention of public health officials. But gold mining has declined dramatically since the late 1980s, accompanying the worldwide plunge in gold prices. At a little under $300 an ounce, gold prices are currently a third of those prevailing during the gold rush to Amazonia in the 1980s. A preoccupation with mercury contamination could blind officials to the "sleeper" issue of potentially dangerous levels of insecticides in widely consumed fish species. Many fish in the Amazon migrate, so if they become contaminated with pesticides along the middle Amazon, consumers along tributaries could also suffer the consequences.

In addition to fish, insecticides could contaminate game animals, rendering them unsafe to eat, or decimate ducks, turtles, and capybaras captured for food. Other environmental issues surrounding pesticide use on the Amazon floodplain include the destruction of pollinators, natural enemies of crop plants, and loss of soil microorganisms involved in nutrient cycling.

One way to eliminate, or at least reduce, insecticide applications would be to deploy pest-resistant varieties of vegetables. At the moment, farmers are purchasing seed of tomatoes and leafy vegetables produced in central and southern Brazil as well as abroad because no home-grown varieties have been developed that withstand local environmental pressures. Two agricultural supply stores in Santarém in 1994 were selling tomato seeds produced in temperate regions of Brazil as well as Europe and the United States. At the Sanagro store in Santarém, for example, seeds of the Santa Cruz variety of tomato, one of the popular varieties for growers on the floodplain and uplands alike, were produced by Agroceres in Minas Gerais, in south-central Brazil. Another variety of tomato on sale in the store, Manglobe Extra, was produced in Denmark. At Agrofort, another agricultural supply store, seeds of the Rutgers variety of tomato were being marketed under the Isla brand from seed imported from the United States via Porto Alegre in southern Brazil. The seeds of several other tomato varieties grown by floodplain farmers, such as Santa Clara and Rio Fuego, are also purchased in stores. Although about half a dozen varieties of tomato are planted on the Amazon floodplain, all of them require chemical protection, particularly if they are grown on a relatively large scale. Pest pressure is likely more severe because farmers generally plant tomatoes in monocultural plots with only one or two varieties.

Vegetable growers in Amazonia would undoubtedly benefit from more research on varieties adapted to the intense onslaught of diseases and pests characteristic of the humid tropics. A broad-based research agenda would include integrated pest management that emphasizes the deployment of resistant varieties, biocontrol of pests and diseases, and improved agronomic practices, such as crop rotation. At present, little if any research is being conducted on developing varieties of vegetables adapted to the ecological and cultural realities of Amazonia.

Modern and Traditional Approaches to Intensifying Cereal Production

The intensification of cereal production is taking place in both modern and traditional forms, as evidenced by two crops: maize and rice.

Maize

In precontact times, maize was made into a fermented drink along the Amazon, and may have been prepared in a variety of dishes. But for the last several hundred years, maize has lingered as a relatively minor crop, grown mostly to feed chickens and pigs. Fresh maize is usually roasted, rather than boiled as in North America. Dried maize is ground into a coarse meal, sweetened with sugar, and baked to make *pomonha*, similar in texture to the Mexican tamale. Pomonha is eaten as a snack food, and in urban areas vendors carry it on trays in search of customers. *Canjica* is a smooth-textured dessert prepared by grating maize and adding sugar and cinnamon. The cornstarch sets the dessert to the consistency of thick custard. Both canjica and pomonha are prepared from maize grown on both uplands and the floodplain, and are consumed in rural and urban areas.

Maize is also consumed as popcorn, most commonly in urban areas. The hard-kerneled varieties of maize used to make popcorn are not grown on the floodplain, however. Popcorn is a rare treat in villages along the Amazon floodplain, where it is usually purchased from small vending carts during festivals. At Arapixuna, for example, popcorn is eagerly purchased by young and old alike during the Círio festival in late July; customers can pay a little extra to have their popcorn topped with ketchup and mayonnaise.

The historically limited use of maize in the Brazilian Amazon has acted as a brake to innovation and intensification. Now shifting markets are changing the picture. As income levels in Brazil rise, so too is the consumption of meat, especially chicken. Poultry is one of the cheaper forms of meat in urban markets, and maize is the main component of chicken feed in commercial operations.

Several ranchers in the Santarém area have begun experimenting with more intensive techniques for growing maize. Three main forces are driving this move. First, local and regional investors are looking for other options now that gold mining profits are down. Second, investors from central and southern Brazil are buying properties on the *várzea* and seeking ways to develop them. And finally, local market prospects for maize have improved now that Santarém has a sizable poultry farm.

The main poultry operation in Santarém, Pioneira, has a processing plant in town and a farm along a side road at km 28 of the Santarém-Rurópolis highway. Owned by VARIG, Brazil's largest airline, the farm feeds fifteen tons of maize a day to its chickens, which are processed at the rate of one hundred tons a week. Pioneira's frozen chickens are marketed from Manaus to Macapá. Almost all the maize for the farm is purchased in Goiás and Mato Grosso and reaches Santarém along either of two routes: by road to Porto Velho, and then by barge down the Madeira and Amazon rivers; or by road to Belém, and then by barge up the Amazon.

Pioneira would prefer to buy maize locally to avoid the appreciable transportation costs of bringing maize 1,500 to 2,500 kilometers from other parts of Brazil. And, because dealers demand cash up front, rather than upon delivery, money is tied up during the two weeks it takes the maize to reach Santarém. If owners could buy maize locally, chickens would be converting the maize to meat within a day or two. Finally, Pioneira plans to expand operations in order to process 130 tons of chicken a week, and the managers would like to diversify their sources of feed.

To help defray costs, Pioneira may soon market at least some of the chicken manure as a local mix in cattle feed, thus maximizing the food value of its purchased maize. A rancher near Souré on Marajó Island, who has diversified his operation with various livestock and crops, already mixes chicken manure with cattle feed. The health implications of this practice warrant scrutiny.

In response to the sizable market for maize in Santarém, the 5,000 hectare São Sebastião ranch on Ituqui Island some thirty kilometers down the Amazon tried to diversify into mechanized and irrigated maize production. After floodwaters receded in 1992, seventy hectares were set aside on a high bank overlooking Paraná do Ituqui for intensive production on the ranch, which contains a herd of 2,000 cattle. The friable, well-drained soil was first plowed with a tractor and then irrigated with silty river water drawn by a diesel-powered pump on a barge moored to the bank. Small amounts of potassium fertilizer were applied, but no insecticides or herbicides were used. The tractor and irrigation equipment were purchased with fiscal incentives provided by the FNO (Fundo Constitucional de Financiamento do Norte) program, which is administered through the state-owned Banco da Amazônia. Overall the diversification effort received only modest support from the credit system.

Expectations were high. The owner of the São Sebastião ranch was advised by an agronomist he had employed that maize yields would be around five tons per hectare, about five times typical yields on uplands in Amazonia. But the harvest, which lasted from late 1992 into early 1993, produced only three tons per hectare, about the same as yields achieved by small farmers on the *várzea* without mechanization, irrigation, or the use of fertilizers.

Erratic irrigation was one of the reasons for the disappointing yield. The owner of the ranch lives in Belém, and the caretaker apparently had little experience with irrigating crops. Another reason for the low yield was that an inappropriate variety was selected. An open-pollinated variety, BR 5107, was chosen since this medium-statured maize withstands the strong summer breezes on the *várzea*. Resistance to toppling is an important consideration when relatively large areas without windbreaks are planted to maize, as was the case on the ranch. Seed of BR 5107 was also readily available from a long-standing co-op CIRAMA (Cooperativa Integral de Reforma Agrária de Monte Alegre), in nearby Monte Alegre. BR 5107 was developed in the early 1980s by EMBRAPA, the Brazilian agricultural research system, and released by EMATER (Empresa de Assistência Técnica e Extensão Rural), the extension service. While BR 5107 withstood winds, it suffered from weed competition, particularly from quiaboarana, malvorana, and lavender-flowered algodãozinho. In 1993, the same variety was planted again, this time using herbicides to control the prolific weeds. But costs were prohibitive in relation to yield gain, and the owner gave up cultivating maize. Some have argued that semi-dwarf maize varieties should be planted on the fertile soils of the Amazon floodplain, but the experience at São Sebastião ranch indicates that such an approach may not prove cost effective.

Maize grown by small farmers along the Amazon is generally tall in order to outpace weeds and ensure a harvest in the event the flood comes early. Some farmers have successfully intensified production by double cropping. The first crop is grown immediately after the floodwaters withdraw, from August to December; this cycle requires irrigation because takes place during the dry season. The second crop does not need watering because rains begin in December or January, but the farmer runs the risk of losing the crop as the waters begin rising in January. Irrigation does not lead to salinization because the annual floods wash away any salts on the soil surface that result from evaporation. Furthermore, irrigation many enhance the fertility of the soil by depositing fresh silt and suspended nutrients. Seed is typically stored under eaves or in rafters until the next planting cycle (figure 5.7). In a good year, farmers can obtain about six tons of maize a year by double cropping. Although maize is usually grown as a monocrop, one farmer along Furo do Ituqui has interplanted his small maize plot with squash.

Traditional maize varieties grown on the floodplain do not rival the height of those observed in parts of Guatemala, where a traveler in the last century noted a variety soaring five meters high.[3] Traditional varieties grown on the floodplain are generally between two and three meters high. Most farmers are willing to trade some yield for height. But some are willing to gamble. In 1996, one small farmer who lives in Silencio, a village along Igarapé do Mamauru near Obidos, planted the Cavalo variety of maize on his property. He chose Cavalo in spite of its relatively short stature because it has exceptionally large cobs and produces well on floodplain soils; unfortunately, he lost two of his five acres of maize to the flood that year. Similarly, some farmers at Piracauera occasionally plant another medium-statured maize, Baixinco, because a larger proportion of the plant is harvestable seed, rather than long stems and leaves typical of taller varieties.

Figure 5.7. Maize "seedcorn" that has been kept dry under the eaves of a house during the high-water season. The maize will be planted in a few weeks as floodwaters recede. Ilha Tapará near Santarém, Pará, September 1993.

Some fifty kilometers upstream from the São Sebastião property on the island of Piracauera do Baixo, the medium-sized Jardim ranch is also experimenting with maize production, but on a smaller scale and using fewer inputs. In 1993, the 500-hectare ranch planted twenty hectares to Abrião, an open-pollinated variety. The owner of the ranch, a gold dealer from Goiás, used his own financial resources to have the land plowed. The tall maize was planted on 20 January 1993 for harvesting in mid-April; no irrigation was necessary because the rainy season was under way. No fertilizers, pesticides, or herbicides were employed. Weeds were suppressed by planting the maize close together. The Amazon bamboo rat was the only significant pest. At the time of my visit in March 1993, the maize appeared healthy and the caretaker estimated that the imminent harvest would yield four tons per hectare. In the past, herds of capybara raided maize fields on the floodplain, but hunters have decimated these enormous rodents along the middle Amazon and they no longer pose a significant threat to the crop.

The cost of mechanization can pay for itself by allowing a larger area to be prepared for planting, especially if double cropping is practiced. One farmer estimated that it takes about seventeen man-days per hectare to manually clear weeds after the waters subside, particularly if clumps of patazana have established themselves. Only a few floodplain farmers have access to tractors; smaller farmers are sometimes able to obtain the use of one from a municipal government, provided that they are willing to cover fuel costs. Poorer families interested in expanding production sometimes pool resources to purchase a diesel-powered pump. Such measures make sense, and could help boost incomes of local farmers if modest amounts of credit were made available for the purchase or lease of such machinery.

Rice

Attempts to intensify rice production on the floodplain using machinery and other intensive measures have also met with mixed results. Rice has been grown in the Amazon since the late 1700s, especially along the lower Amazon, Tocantins, and Rio Branco. Today, most of the rice grown along the Amazon is confined to the estuarine area, where small farmers usually obtain yields of about three tons per hectare. More intensive production is found along parts of the Rio Branco in Roraima, where farmers from southern Brazil have deployed tractors and irrigation since the late 1980s. By 1986, some 3,000 hectares of floodplain in the Rio Branco watershed were in intensive rice cultivation, providing about one-third of Boa Vista's demand for the commodity. For the most part, though, rice is an upland, rainfed crop in Amazonia, where yields usually average one ton per hectare, and only one crop can be grown a year. Most of the rice consumed in the Brazilian Amazon comes from central and southern Brazil, particularly the state of Goiás.

One rancher on Piracauera Island in the middle Amazon attempted to irrigate a thirty-hectare rice field in 1992 by first mechanically preparing the land. Apparently the field was not level, as banks tend to slope gently to the center of islands, so it was not possible to maintain a uniform water level for the growing rice plants. Yields were disappointing, and in 1993 the rancher switched to beans and vegetables in his failed rice field.

The most ambitious attempt to intensify rice production on the Amazon floodplain took place at Jari during the 1970s, while the multifaceted operation involving pulp production, kaolin mining, and livestock raising was still under the helm of the late Daniel Ludwig. Ludwig, an American billionaire and industrial magnate, was driven in his later years by a conviction that the Amazon could be "conquered." Plans called for diking and leveling 12,700 hectares of floodplain for irrigated rice production. An intricate system of pumps was installed, and light aircraft were used for seeding and pest control. A fleet of large combine harvesters were purchased to reap the crop. While yields were reportedly high, in the range of five to ten tons per hectare, outlays were exorbitant in relation to the market value of the product. The cost of operating the diesel pumps for controlling water levels was particularly onerous. The project was conceived in an era of cheap gasoline, and implemented during a period of steeply climbing petroleum prices.

By the late 1970s, Jari's rice operation was losing between $8 and $10 million per year. Not surprisingly, rice cultivation was discontinued in 1988 when 4,150 hectares were in production—about a third of the projected area for the crop. Jari now grazes a large herd of water buffalo on 50,000 hectares of floodplain pastures instead. For a while, a Brazilian entrepreneur based in Belém experimented with raising pirarucu, a high-value fish, in the network of canals in the former rice area, but this venture has folded. One Brazilian agronomist remarked wistfully on the outcome of Jari's efforts to intensify rice production: "Whoever works on poor man's crops will be poor" (*"Quem trabalha com cultura do pobre, fica pobre"*).

Not all entrepreneurs have given up on "poor man's" crops. For efficient producers, there is money to be made in growing food crops for the market. In

Santarém, a cereal merchant, Francisco Quincó, has been experimenting with mechanized rice production on the floodplain near Alenquer since the early 1990s. Interestingly, his modest business has made the most promising strides with intensified rice production. Francisco's father, José, came to Santarém from drought-plagued Ceará and worked his way up from a wage laborer to a merchant who bought rice from small farms on the uplands, then dried and de-husked it before placing it on the market. Eventually he founded COPAMAZON (Cooperativa de Produtores e Pesquisadores Agrícolas do Médio e Baixo Amazonas), a profitable business in which he still plays an executive role, even though he remains illiterate; "too busy to learn to read and write," as he explained. His son, Francisco, is the managing director of COPAMAZON, which provides him with capital to sustain the experimental phase of rice growing, which was set up as both a yield trial and a moneymaking proposition.

In 1994, the company planted 800 hectares to rice along a bank on the floodplain near Alenquer. The field was plowed by tractor and fertilized at the relatively modest rate of seventy kilograms of NPK (nitrogen, phosphorus, and potash) per hectare. Although the floods came early that year, and some of the rice was lost, yields averaged six tons per hectare, about twice the average on the Amazon floodplain under traditional management. Seven varieties were planted (Orixca-1, IAC 409, IAC 412, IAC 414, IAC 416, Xuí, and Taí), ranging from modern to traditional. A seventy-day rice (IAC 416) was included, a sensible strategy in view of the potential for early floods. Eventually, Francisco will likely narrow the number of varieties to two or three, but he has wisely avoided the potentially discouraging approach of trying only 1 variety for the entire operation.

Manioc, Amazonia's Daily Bread

Long cultivated in the Amazon, perhaps even domesticated there, manioc has always been an important food crop in the region. Farmers have had ample time to select a large number of cultivars to suit their cultural and local ecological conditions. The total number of bitter and sweet manioc cultivars in the Brazilian Amazon most likely exceeds a thousand, especially if the unique varieties cultivated by surviving indigenous groups are taken into consideration. For example, the Kuikuru of the Upper Xingu maintain some fifty manioc varieties, while the Jívaro of the Ecuadorian Amazon can name one hundred cultivars. Four villages of the Tukano in northwest Amazonia have 123 varieties, and the Amuesha of the Palacazu Valley in the Peruvian Amazon cultivate more than 200 varieties.

Manioc is a bush with starchy tubers that is propagated by stem cuttings. Technically a perennial, it can also be considered an annual crop because it is sometimes harvested within six months. Two basic types are grown: the "bitter" form, used for making flour that can be stored for months, and "sweet," which is peeled and boiled. The bitter kind contains varying amounts of prussic acid and is toxic; that is why it is rarely bothered to any extent by predators and produces a reliable yield. The process of making flour eliminates the poison, rendering it fit for human consumption.

Of the seventy-nine varieties of bitter manioc I have noted in fields in the Brazilian Amazon, seventeen occur on the floodplains of the middle and lower

Amazon (appendix C). Of twenty-two cultivars of sweet manioc observed, nine are grown on the floodplain (appendix D). The annual flood along the Amazon acts like an ecological sieve; only precocious cultivars of manioc can be grown with assurance on the floodplain. On the uplands, in contrast, many cultivars can be harvested in stages up to twenty-four months after planting. Most of the manioc on the uplands is harvested in the rainy season because it is easier to pull the roots out of the ground. Farmers often plant as they harvest and cuttings have a better chance of taking root during the wetter months.

Of the seventeen cultivars of bitter manioc recorded on the floodplain, ten are grown in upland areas. At planting time on the *várzea*, some farmers gather stakes of bitter and sweet manioc from their upland fields and take them to the floodplain (figure 5.8) where they are chopped into hand-length segments and inserted into the recently exposed soil (figures 5.9, 5.10). For varieties that appear to be restricted to the floodplain, such as Dorotea and Piraíba (a pimelodid catfish), the tubers are harvested as the waters rise and some of the stakes are either stacked on high banks (figure 5.11) or on makeshift platforms, as observed on Careiro Island in 1972. Stakes of some varieties that are grown on both the uplands and the floodplain, such as Amarelinha (little yellow one), Flor de Boi (cattle flower), and Gordura (fat), are also sometimes stored on the *várzea* during high water.

Of the nine cultivars of sweet manioc found on the Amazon floodplain, two appear to be grown only on the *várzea*. Even if the stakes are partially submerged for a month or more, they are still viable. As would be expected, farmers with no access to upland fields maintain several varieties that have been honed for the floodplain. Manioc yields on the floodplain appear to be high, in spite of the short growing season. Near Tabatinga, for example, yields of thirty-five tons per hectare have been recorded on the higher banks of the Amazon within four months

Figure 5.8. Stakes of the Tartaruga variety of bitter manioc harvested from an upland field for transplanting on the floodplain. Arapixuna, Pará, July 1996.

Figure 5.9. Farmer digging a hole with a hoe to plant manioc on the floodplain. Igarapé Jari near Arapixuna, Pará, August 1993.

Figure 5.10. Farmer and son taking a break from planting manioc on the floodplain. Long manioc stems are in the foreground, and smaller segments ready for planting are by the log upon which the mother and son are seated. The calabash gourd to the left contains drinking water. Igarapé Jari near Arapixuna, Pará, August 1993.

Figure 5.11. Manioc stakes stacked by a floodplain farmhouse. Floodwaters are receding and the stakes will soon be cut up into segments and planted. The farmhouse sports a television antenna because it is relatively close to Óbidos, Pará. June 1994.

of planting. The mix of cultivars that farmers deploy on the floodplain is thus partly a function of whether they maintain fields in upland areas.

Manioc is typically grown as a monocrop, but farmers frequently interplant two or three varieties on the floodplain near Santarém. Along Surubim-Açu, a side-channel of the Amazon, for example, one farmer intercrops three bitter varieties—Amarelinha, Flor de boi, and Roxinha (little red one). In some upland areas of Amazonia, an even greater diversity of cultivars can be found. On the outskirts of Vila Socorro by Lago Grande da Franca near Santarém, I found nine intercropped varieties of bitter manioc—Acarí, (a loricarid catfish), Achada (found), Caratinga (a fish in the cichlid family), Nambu (tinamou), Perereca, Preta (black), Roxinha, São José, and Zucuda—in a single field on an upland bluff close to the Amazon floodplain. A couple of the varieties encountered in that field are also grown on the Amazon floodplain. Sometimes the varieties are kept in separate blocks or zones within a field, in other cases they appear to be interplanted in random fashion.

In the Santarém area, farmers offered two reasons why they often interplant several manioc varieties in the same field. When starting a new manioc field, a farmer may not have enough cuttings from his own stock; he therefore borrows or barters for planting material from neighbors. By the time the farmer has solicited material from neighbors, he usually has at least two varieties for planting. Another reason for planting more than one variety is to improve the taste of the flour. Yellow flour is highly desirable from the commercial perspective, but varieties with yellow tubers, such as Burrona, tend to taste slightly bitter. This unpleasant taste can be removed by blending the dough with other varieties, particularly white-tubered Chavita. Another advantage of Chavita is that it is a high yielder (*rende mais*) and produces generous roots (*engrossa mais a batata*).

Additional factors account for the generous repertoire of manioc varieties along the Amazon. Near the Camará River on Marajó, for example, one farmer asserted that he maintains the Paruí variety because it is good for intercropping. Tall Paruí, with its spindly branches and slender leaves, does not cast excessive shade for underlying crops, such as seedlings of fruit trees. Differences in taste and texture also account for some of the diversity of sweet manioc varieties. An added bonus of planting a mosaic of cultivars is that it probably reduces pest and disease pressure.

Some indigenous groups intercrop far more varieties of manioc than is commonly found in fields prepared by farmers more integrated with national society. In the Colombian Amazon, the Tukano interplant as many as forty-eight cultivars, while the Tiriyó in northern Amazonia cultivate as many as twenty manioc varieties in a field. The Aguarana, a Jívaro group along the upper Marañon in the Peruvian Amazon, intercrop an average of twelve cultivars.

Cultivar names of manioc in the Amazon reflect the intimate contact people have with nature. Many of the varieties are named after animals, such as Anta (tapir) and Tartaruga (the giant river turtle). Birds also feature in a number of cultivar names, including Arara amarela (the yellow macaw), Bemtevi (kiskadee flycatcher), Juriti (ground dove), and Nambu (tinamou). Fish names are ascribed to some bitter varieties, including Acarí, Caratinga, Piraíba, and Tambaqui (a fruit-eating characid). Several bitter manioc cultivars are named after fruits, both wild and domesticated, including Abacatinho (small avocado), Açaí, Bacuri, Cacao, Inajá (an upland palm with oily fruits), Mamão (papaya), Pororoca (a wild tree of uplands in Central Amazonia with edible fruits), and Tucumã.

Some of the cultivar names are probably synonyms, particularly names implying diminutives, such as Nambu and Nambuzinho, or Inajá and Inajázinho. Various genetic fingerprinting techniques would need to be applied to samples to sort out the distinctiveness of the various cultivars. Even if some of the cultivars in my sample prove to be duplicates, the actual number of manioc varieties grown in the Santarém area is likely to be higher than my field surveys indicate.

The preponderance of bitter manioc varieties is a reflection of the fact that most of the crop is converted to flour, or prepared into a storable pancake called *beijú*. Farmers prepare manioc flour on the Amazon floodplain for domestic consumption and for growing urban markets. The manufacture of flour is especially labor intensive on the *várzea* since traditional methods of peeling, grating and squeezing the tubers are mostly used. Manually operated graters, instead of gasoline or diesel-powered machines, are more common. Also, the sleeve-like tipití (figure 5.12) is typically used to squeeze the dough, rather than various box presses typically used by settlers from southern and central Brazil on the uplands. In both upland and floodplain areas, neighbors or relatives sometimes collaborate when making manioc flour. When family labor is insufficient, arrangements are made with workers to keep half of the manioc flour in lieu of wages.

Two by-products from the making of manioc flour are widely consumed in rural and urban areas of the Amazon: tucupi sauce and tapioca. Tucupi sauce,

Figure 5.12. A farmer sitting on a pole to tighten a tipití press so that the juice is squeezed out of the grated manioc. The tipití is fashioned from the jacitara palm gathered in upland forest. The juice is collecting in the spathe of an inajá palm, also obtained from upland forests. Murumuru near Santarém, Pará, July 1996.

prepared from the liquid squeezed from manioc dough, is a savory addition to a variety of regional dishes, including fish stews, soup, and boiled duck (*pato no tucupi*). Tucupi is safe to consume once the liquid has been boiled. One dish featuring tucupi sauce as a major ingredient, called *mujica*, employs the remains of roasted fish with olive oil, manioc flour, and assorted spices. Small, dried shrimp is sometimes added to mujica. Pirarucu in tucupi sauce (*miúdo de pirarucu no tucupi*) is another example of home cooking with the sauce, popular in the Santarém area. Pirarucu is an expensive fish so only the "parts" (*miúdo*) are used in the recipe, which includes garlic and the tangy leaves of jambu.

Tucupi sauce is also an ingredient in *tacacá*, a soup of dried and salted shrimp, semiliquid tapioca, and the leaves and flowers of jambu. Yellow-flowered jambu is cultivated, but also occurs as a weed in houseyards; it contains a compound that causes a tingling sensation on the lips. Tacacá stands are a familiar sight on sidewalks in towns and cities in Amapá, Pará, and to a lesser extent in Amazonas. *Tacacá* soup is served in blackened calabash gourds, and customers either stand or wait for a spot to open up on the narrow wooden bench supplied by some vendors (figure 5.13). Housewives also prepare the soup for their families; one or two bowls of it can substitute for dinner.

Farmers prepare three main types of tucupi sauce for domestic consumption and sale in towns and cities: plain, yellow-colored, and orange-tinted. The yellow-colored sauce contains the mild, aromatic fruits of a variety of capsicum pepper called *pimenta de cheiro*. The round, yellow fruits impart a distinctive color to the sauce after storage in a bottle for several days. The orange-colored sauce is

Figure 5.13. A tacacá stand in Macapá, Amapá, December 1994.

obtained by adding a few pungent fruits of another capsicum pepper, *pimenta malagueta*. Its fiery red fruit is fair warning of its potency; only a few are added to the bottle and left in the sun to "cure" for a few days. Both varieties of capsicum pepper are grown in large tin cans or elevated spice beds in backyards.

An unusual use for manioc juice was observed in Arapixuna near Santarém. In an effort to control leaf-cutter ants, a farmer poured the liquid squeezed from manioc dough into their nests. Leaf-cutter ants are notorious for damaging crops and sometimes force farmers to abandon fields; they are considered the most serious pest of manioc in the Brazilian Amazon. It is ironic, then, that a by-product from the crop itself is used to combat the pest. Manioc juice does not appear to kill the ants outright. Rather, the farmer explained that it "annoys" the ants, and flushes them from their nest; he noted wryly that they probably head for his neighbor's manioc field. People in Amazonia and other parts of Brazil have been devising ways to combat leaf-cutter ants for a long time. In Bahia during the sixteenth century, for example, farmers once placed old manioc leaves around the nests of leaf-cutter ants to distract them from crops.

Tapioca is prepared from the starchy sediment that decants from the juice squeezed from manioc dough. It is served in a variety of ways: as a gummy, translucent paste in tacacá soup; as a chewy crêpe (*beijú de tapioca*); as a porridge (*mingau de tapioca*); and as a "popcorn" (*pipoca de tapioca*). Tapioca crêpes are served at daybreak with coffee, both at home and on the street. The vendor fries the patty quickly, spreads butter or margarine to taste, then rolls up the crêpe and hands it over to the customer in a paper napkin. Some clients request a sprinkling of grated coconut. All family members delight in this filling repast, typically eating perched on a rickety wooden bench supplied by the vendor.

Tapioca porridge is usually served with rice in calabash gourds. Street vendors sell tapioca porridge in the morning in towns along the Amazon, such as Alenquer and Mosqueiro (figure 5.14). Crunchy tapioca popcorn, prepared by toasting tapioca beads on a griddle so that they puff up, is the poor man's answer to regular popcorn. Children in villages such as Arapixuna consider tapioca popcorn a treat. Tapioca pudding can be found in the dairy sections of supermarkets throughout the United States, where it is eaten as a dessert. Children, especially when they are colicky, find tapioca pudding easy to digest.

Squashes, A Riot of Color and Form

Squashes produce abundant harvests on the fertile banks of the Amazon and its side-channels, where they are grown for domestic consumption and urban markets (figure 5.15). A surprising diversity of shapes, colors, and textures characterizes the squashes, ranging from giant crook-necked types to forms resembling small cartwheels (figure 5.16). In Alenquer in March 1993, I observed 8 distinct varieties—Caboclo (peasant), Cacao, Collor, Comum (common), Leite (milk), Pescoço (neck), Peito de moça (girl's chest), Pneu (tire)—all from the floodplain. The floodplain may prove to be an overlooked area for genetic diversity of squashes, and new forms appear to be arising. For example, the vari-

Figure 5.14. Street vendor pouring tapioca porridge into a calabash gourd for a customer's breakfast. Mosqueiro, Pará, June 1994.

Figure 5.15. Several squash varieties for sale in a street market in Alenquer, Pará, March 1993.

Figure 5.16. A wheel-shaped squash, grown on the floodplain, for sale in a street market. Several varieties can be seen in the background. Mats of the junco reed, used for covering produce, can be seen on the right. Macapá, Amapá, December 1994.

ety Collor, which I saw in the Alenquer market, is named after Collor, a recent president of Brazil.

Squashes are sometimes grown as an intercrop with maize or manioc. The most important species appear to be *Cucurbita moschata* and the closely related *C. maxima*, although they are often difficult to distinguish, especially after they have been harvested. Sometimes called squash gourd in English, *C. maxima* is generally considered to have only two varieties: winter squash (var. *maxima*) and turban squash (var. *turbaniformis*). But *C. maxima* is indigenous to South America, and may even be native to the Amazon Basin, where more varieties surely await to be uncovered. The other important squash species on the *várzea*, *C. moschata*, is native to Central America and has likely been in Amazonia long enough for locals to have selected unique forms.

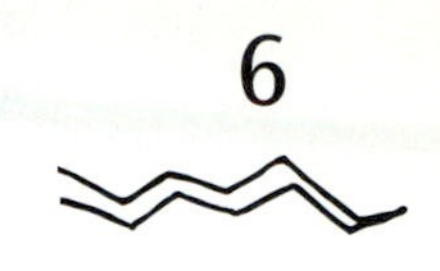

A Harvest for All Seasons

Tree farming is one of the most overlooked avenues for promoting agricultural development on the Amazon floodplain and for helping to restore some ecological balance to current land use. Some argue that the Amazon's sweeping annual floods make it a less than propitious setting for agriculture, particularly for tree crops. But the "vocation" of the floodplain may indeed be in perennials, rather than the pasture that prevails today. Growing perennials is less risky than cultivating annuals in periodically inundated habitats, and many trees and bushes are adapted to the Amazon floodplain.

Trees once dominated Amazon landscapes, from the high levees bordering the main channel of the river to margins of the myriad lakes and somber backswamps. Today, in contrast, the floodplain probably has fewer trees than at any time in its long history. Even at the height of human population density in precontact times, much of the region was blanketed in forest albeit often altered by human activities, because indigenous people did not raise livestock. When Friar Carvajal and other early explorers reported seeing "forest" along the Amazon they were probably looking at multispecies orchards planted by indigenous groups, at least in some cases. But after the collapse of indigenous civilizations, the forest reasserted itself. Little significant clearing took place in the colonial period, when economic activities focused on extraction of forest products.

The last two or three decades, however, have witnessed a dramatic withdrawal of floodplain forests, particularly in the middle Amazon, primarily as the result of cattle and water buffalo raising. Ranching is likely to remain a major economic activity in the region, at least in the near term. Various options, such as cash cropping with perennials, could be pursued to slow or arrest the rising tide of deforestation. While punitive approaches to checking the expansion of livestock raising are unlikely to work, positive alternatives, particularly agroforestry, show promise.

Agroforestry for the Amazon floodplain is not a new idea. Four main agroforestry systems have evolved on the middle and lower Amazon: home gardens; managed fallows; enrichment planting in forests; and polyculture in cacao and rubber stands. Over time, farmers have planted fruit trees in old cacao and rubber stands, such as on Careiro Island near Manaus. But many of these multipurpose plantations are now losing ground to pasture.

This chapter focuses on innovative approaches to commercial agroforestry and the role of home gardens in the agricultural development of the floodplain. For example, many of the forests in the Amazon estuary have been transformed with economic plants, especially açaí. In some areas along the middle floodplain, such as at Murumuru near Santarém, farmers have enriched the forest with economic species such as rubber and açaí. Such practices can provide valuable clues as to how to restore degraded areas.

The Pivotal Role of Home Gardens

Commonly perceived as backyard collections of plants for domestic use, home gardens may seem an unlikely source of ideas for commercial agroforestry. Yet these gardens are rich islands of agrobiodiversity. Numerous wild species from the forest and other environments along the Amazon are being domesticated in backyards for a wide variety of purposes, and some of these species could be major crops in the future. Home gardens are also low-cost arenas for trying out new crops from other regions. If the introduced plants perform well, they may be cultivated on a larger scale. In addition, the partial shade of home gardens is ideal for establishing seedlings, such as cacao and açaí palm, for eventual transplanting in groves. Finally, home gardens are important depositories of varieties of crops adapted to occasional flooding, such as mango, cupuaçu, banana, cashew, and guava (figure 6.1).

Although home gardens cover relatively small areas, typically less than 3 hectares, they are the most species-diverse cropping systems in Amazonia. A survey of twenty-two home gardens on the floodplain of the middle and lower Amazon revealed ninety-three plant species (appendix E). If flowers, vegetables, herbs, and spices are considered, it is safe to assume that several hundred plant species are cultivated or tended in home gardens in that area. Up to two dozen perennial species can be found in a single home garden on the floodplain. A single dooryard garden on Ilha das Onças in the Amazon estuary contained sixty-eight useful perennial and herbaceous plants. Home gardens, with their extraordinary species richness, are a valuable source of promising crops for more widespread planting.

Not only is the diversity of species in home gardens impressive, but for some crops, numerous varieties are also maintained. For example, at least sixteen cultivars of bananas and plantains are found in home gardens on the floodplain below Manaus (table 6.1). And in a single home garden on Carmo Island, five varieties of banana were found. Although some of the cultivar names may be synonyms, it is likely that several dozen varieties of bananas and plantains are cultivated in home gardens along the entire floodplain. Locals appreciate the differences in taste

Figure 6.1. Harvesting guava in a home garden at high water. The banana trees are also producing fruit. The yard is under about a meter of water. Paraná Cachoeri near Oriximiná, Pará, June 1994.

and texture between varieties. The standard Cavendish banana sold in supermarkets in North America and Europe is relatively bland compared to the dense, sweet flesh of some traditional cultivars. While the diversity of banana cultivars on the floodplain pales in comparison to that found in Southeast Asia where the crop originated, the mutants selected by farmers in the American tropics, including Amazonia, are creating secondary centers of diversity for the crop. Several varieties appear to have been selected to tolerate flooding. For example, seven of the sixteen cultivars encountered on the floodplain are not recorded from uplands. Only four of the thirty-seven banana and plantain cultivars that I have recorded in home gardens and fields in the Brazilian Amazon are listed in an authoritative global list of cultivar names for the crop. Clearly, much work remains to be done in sorting out the number and uniqueness of banana cultivars in Amazonia.

Little significant difference is found in species diversity between home gardens on the uplands and the floodplain. I found seventy-seven species in thirty-three home gardens in upland areas of the Brazilian Amazon. But species composition differs markedly. The annual flood along the Amazon acts as a filter: of the ninety-three species I found in floodplain home gardens, less than half are also found in backyards on the uplands.

Many species, therefore, are unique to home gardens on the floodplain. This is especially so for the half dozen or so species of fish-bait trees and bushes. Since homes are often located on the higher parts of the floodplain, some farmers opt to

Table 6.1. Some banana/plantain cultivars maintained by farmers in the Brazilian Amazon

	Amazon Floodplain		Uplands	
Name(s)	Home Garden	Field	Home Garden	Field
Baié				+
Baixinha, cambota	+			
Baixota				+
Branca	+	+	+	+
Casada	+		+	+
Casca verde	+	+	+	
Catavé				+
Catuá				+
Chifre de boi*	+			+
Comprida*				+
Costela de vaca			+	
Couruda				+
Degenerada	+		+	
Fufa	+			
Grande*	+		+	+
Grossa	+			
Inajá	+	+	+	+
Jangauebí				+
Landeza			+	
Maçá			+	+
Mação	+			
Maçarola	+			
Maraquinha	+			
Maraquita			+	
Mysore			+	
Nanica				+
Nanicão				+
Pacova, pacovão, pacovi	+	+		+
Perua, Peruara			+	+
Piraíba				+
Prata	+	+	+	+
Roxa			+	+
São Tomé amarela				+
São Tomé roxa				+
Sapo	+			
Tres Penca				+
Tucum				+

*plantain

Notes: According to one farmer, *pacovão* and *chifre de boi* are synonyms; other cultivar names may also be synonymous.

Home gardens on the floodplain were sampled from Careiro Island near Manaus to Afuá on Marajó Island. On uplands, gardens were inventoried in widely scattered locations in Pará, near Machadinho in Rondônia, and in the vicinity of Figueirópolis, Mato Grosso.

grow a few crops that cannot tolerate flooding. Avocados, for example, fetch a handsome price in urban markets, so farmers plant them in their home gardens, even though an exceptional flood every dozen years or so may kill the trees. But varieties of crops normally associated with uplands, such as sweet orange, have also been selected for tolerance to flooding (figure 6.2).

Floodplain residents accumulate a diverse array of plants around their homes and use them for a wide variety of purposes (table 6.2). The trees most commonly planted are fruit bearers, used for fresh fruit or juice. Ingá capuchino, for example, produces small pods that resemble those of broad beans (figure 6.3), and is one of several species of *Inga* that are planted or gathered in the wild. It is probably named after the Capuchin monks that operated in the area during the colonial period. Some three dozen species are cultivated for fresh fruit alone along the middle Amazon, a pattern that also prevails along the upper reaches of the river. Many of the trees found in home gardens are exotics. Among them are the mango and banana, the two most commonly encountered trees in floodplain home gardens. Home gardens also serve as a pantry, medicine chest, and hardware store. The calabash tree, for example, supplies bowls for bailing out canoes, saving seed, storing water, and for stashing food items (figures 6.4, 6.5).

Some sixteen species of forest trees are deliberately planted in home gardens along the middle Amazon. These protodomesticates range from fruit trees such as açaí, bacaba, bacuri, marimari, tucumã, yellow mombim, and various species of *Inga*, to fish baits such as catauarí and socoró. Other seedlings uprooted in the forest for transplanting in home gardens include sapucaia for its nuts,

Figure 6.2. Harvesting sweet oranges in a home garden along the banks of the lower Maracá River, Amapá, May 1996. The river receives silt-laden waters from the Amazon at high tide. During the high-water season, this orange tree is flooded twice daily by tides, but it is not inundated during the dry season.

Table 6.2. Some local uses of plants grown in home gardens on the Amazon floodplain from Manaus to Marajó Island

Use	Plants
Alcoholic drink	Genipap
Bowl	Calabash gourd
Broom	Vassoura
Caulking	Cotton
Condiment/food flavorer	Lime, capsicum pepper
Hot beverage	Cacao, arabica coffee
Fish bait	Araçá, baianeiro, catauarí, curumi, socoró, tarumã, uruá
Food colorant	Annatto
Fruit	Açaí, ambarella, avocado, bacaba, bacuri, banana, biribá, breadfruit, buriti, cacao, cashew, cutitiribá, genipap, guava, ingá baú, ingá capuchino, ingá cipó, ingá grande, jambolan, Malay apple, mango, marimari, muruci, papaya, peach palm, pineapple, pitomba, pojó, sapodilla, socoró, sweet orange, sweetsop, tangerine, tarumã, tucumã, watermelon
Fuelwood	Boteiro, cecropia, curumi, marinemera,* meracruera,* mulato wood, paricá da várzea, pojó, sapupira
Juice	Buriti, carambola, cashew, coconut, cupuaçu, passionfruit, soursop, sugarcane, tamarind, yellow mombim
Lamp wick	Cotton
Livestock feed	Apé (ducks), cannonball tree (pigs, domestic fowl), sapucaia (pigs, chickens); patazana (chickens); uruá (chickens)
Living fence	Yellow mombim
Medicinal	Andirá-uxi, jucá, marisara,* pião branco, pião roxo
Mulch	Munguba (rotten trunks and branches)
Nectar for bees	Tachi*
Nuts	Cashew, sapucaia
Preserve/jam	Surinam cherry
Shade	Andirá-uxi, aruanãzeiro, boteiro, caçeira, carnaúba palm, caxinguba, marinemera,* sapupira
Skewers for roasting fish	Baianeiro
Soap	Pau de Angola
Staple food	Common bean, maize, manioc, runner bean, squash, sweet potato
Timber	Andirá-uxi, mulato wood

*unidentified

Figure 6.3. Harvesting the fruits of ingá capuchino in a floodplain home garden. Several species of *Inga* and its near relatives occur wild in floodplain forests, and some of them are in early stages of domestication in home gardens. Urucurituba near Santarém, Pará, March 1993.

Figure 6.4. Calabash gourd trees in a partially submerged home garden that straddles the interface between uplands and the floodplain. A ripe gourd can be seen in the tree on the right. Calabash gourds are cut in half to make bowls for use in the kitchen and to scoop water out of canoes. Carariacá, Pará, June 1994.

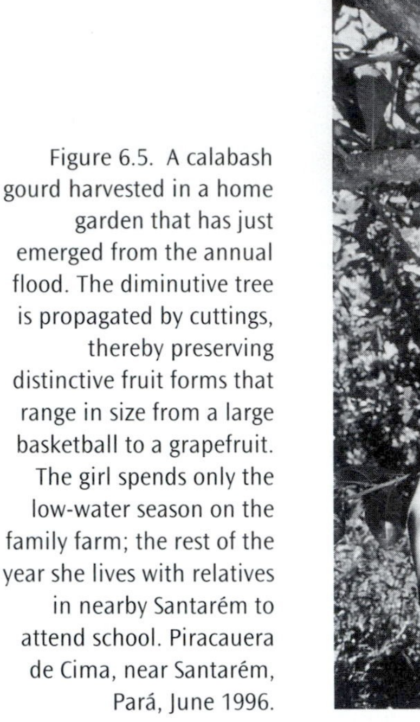

Figure 6.5. A calabash gourd harvested in a home garden that has just emerged from the annual flood. The diminutive tree is propagated by cuttings, thereby preserving distinctive fruit forms that range in size from a large basketball to a grapefruit. The girl spends only the low-water season on the family farm; the rest of the year she lives with relatives in nearby Santarém to attend school. Piracauera de Cima, near Santarém, Pará, June 1996.

the cannonball tree for livestock feed, and andirá-uxi and mulato wood for timber.

The origin of most of the domesticated plants of lowland South America, as well as other parts of the humid tropics, is to be found in home gardens. And women have been at the center of much of this action; tending spontaneous seedlings, dispatching children into the surrounding forest in search of desirable seeds and seedlings, and carefully watering and weeding young plants. The process continues along the Amazon floodplain and in upland locations. Plants are brought along the path to domestication in various ways: by sparing certain trees when a homesite is cleared, by protecting spontaneous seedlings, and by seeking out seeds or seedlings in the forest. At one home on Urucurituba Island, for example, young boys were sent into the forest to collect seedlings of the cannonball tree for planting in the backyard. The family had also uprooted seedlings of sapupira in the forest, and after stabilizing them in pots, were transplanting them for fuelwood.

The tending of spontaneous plants has been called *arboriculture* by the geographer David Harris and *incidental domestication* by David Rindos, an anthropologist.[1] But *incipient domestication* might be a more appropriate term since "arboriculture" simply means raising trees, and "incidental domestication" implies a rather casual attitude toward plants on the threshold of domestication.

In the home gardens sampled along the Amazon, twenty-two spontaneous plant species are generally spared. Several fish-bait trees are deliberately tended when they show up, particularly catauarí, curumi, tarumã, and uruá. Wild fruit trees tended in home gardens include the floodplain bacuri, pitomba (which resemble brown-skinned Muscadine grapes in appearance but not taste), and lychee-like pojó (table 6.2). Seedlings of wild trees useful for timber or fuelwood are encouraged, especially mulato wood, boteiro, and paricá da várzea. Baianeiro is spared because its twigs make convenient skewers for roasting fish. Trees and herbaceous plants allowed to sprout because they provide feed for livestock include urucuri palm, the fruits of which are relished by pigs; patazana, the seeds of which are scattered for chickens; and apé, whose chestnut-sized tubers are much appreciated by ducks. Spontaneous seedlings of munguba are welcome because the seeds provide fish bait, the inner bark is used to string fish, and when the tree eventually dies, the rotting wood is a coveted mulch for vegetable beds. Other trees that are tolerated because they provide shade or are used in folk remedies include aruanãzeiro, caxinguba, and marisara (unidentified), the latter covered in spectacular pink blooms as the floodwaters subside.

A third of the trees and shrubs in home gardens occur wild on the Amazon floodplain (appendix E). Without surrounding forests, home gardens would not boast such a large selection of plants on the verge of domestication. The impressive biodiversity in floodplain forests supplements the diet, generates income, saves money that would otherwise have to be spent on purchasing the items, and provides new crop candidates.

Orchards for a Water Eden

Although the notion of commercial agroforestry for the Amazon floodplain may seem a strange proposition, it is an ancient, if diminished practice. Until the 1950s, cacao and rubber were often grown together and generated some income for farmers along the middle and lower Amazon. A little downstream from Óbidos, for example, a large plantation of cacao (*cacoal*) was intercropped with rubber and banana. Known as Cacoal Imperial, the plantation was started in the seventeenth century by Jesuits. But only in a few locations on the floodplain, such as on Combu Island near Belém, where cacao blends into a cultural forest enriched by many other economic species, is the crop still grown on any significant scale (figure 6.6).

Many of the agroforestry systems on the floodplain, including those involving cacao, have since fallen into disrepair or have vanished. Farmers between Nhamundá and Santarém often cite the unusually severe flood of 1953 as the triggering factor in their decisions to abandon cacao cultivation. But cacao was on its way out as a commercial crop before then; severe floods in 1854, 1859, 1892, 1895, and 1900 had already wiped out many of the plantings between Faro and Alenquer. The gradual abolition of slavery in Brazil in the 1880s spelled the demise for some of the larger plantations along the Amazon and cacao exports from the Brazilian Amazon peaked at 7.5 million kilograms in 1888. Falling prices of cacao and rubber in the 1980s are another reason why cacao is less common and rubber is left untapped. The jute boom of the middle of this century, coupled with the recent expansion of cattle and water buffalo ranching, have laid waste to many intricate and formerly extensive agroforestry plots on the floodplain.

Several farmers pointed to pasture where cacao groves once stood, and one grower remarked wistfully to me that cacao orchards used to be favorite places

Figure 6.6. Cacao pods gathered from a forest enriched with cacao and other economic plants in the Amazon estuary. The baskets were used to carry the pods to the canoe. The farmer will shortly split open the pods with a machete to remove the pulp-covered beans, which will then be fermented and sun-dried on mats. Combu Island near Belém, Pará, December 1994.

for rendezvous during courting because they had relatively cool interiors with closed canopies for discretion. Henry Walter Bates captured the ambiance of a cacao grove along the Amazon in the mid-nineteenth century:

> It was very pleasant to roam in our host's cacoal. The ground was clear of underwood, the trees were about thirty feet in height, and formed a dense shade. Two species of monkey frequented the trees, and I was told committed great depredations when the fruit was ripe.[2]

Gone, too, are the annual cleanup of cacao orchards and the associated party known as *puxirum*. Cacao growers would organize groups of neighbors or estate works to prune the trees and weed, providing liberal quantities of white rum (*cachaça*), brought from Abaetetuba on the lower Tocantins in twenty-liter glass flagons called *frasceiras*. A beef barbecue was also provided, served with rice and manioc flour. In lieu of payment, owners of small groves would reciprocate by assisting in the cleanup of their neighbors' cacao plantings. According to old timers in the Santarém area, the last puxirum was held in the mid-1960s. And the small-scale cachaça stills in the Amazon have been put out of business by industrial-scale distilleries in the Northeast and São Paulo.

Small farmers have adopted agroforestry on a commercial scale in many upland areas of the Brazilian Amazon, but this widespread trend has largely bypassed the *várzea*. The Amazon floodplain has been virtually ignored by development planners for several decades. The push to open roads in interior forests and establish colonization schemes is a major reason why much of the commercial agroforestry is on *terra firme*. Brazil's cacao research and development agency (CEPLAC—Comissão Executiva do Plano da Lavoura Cacaueira), for example, has focused its R&D efforts in Amazonia on the uplands, particularly areas with alfisols, such as near Altamira in Pará and in Rondônia.

Today, rural folk derive a significant livelihood from perennial cropping and agroforestry only in the Amazon estuary area, and in particular close to Belém. On Combu Island, for example, polycultural orchards containing a mix of wild and cultivated species have replaced slash-and-burn agriculture as the mainstay for farmers. But what is the potential for agroforestry for less privileged locations of the floodplain? No single model or agroforestry configuration is appropriate for the entire stretch of the river. Environmental heterogeneity and differing cultural and market conditions preclude any one blueprint for development. If agricultural scientists and extension agents work closely with farmers to analyze socioeconomic constraints and identify market opportunities, it should be possible to promote profitable agroforestry systems on already deforested areas of the floodplain.

A two-pronged strategy would likely reap benefits for rural producers and urban consumers alike. First, agroforestry development should focus on a few crops with secure markets, such as some of the well-known fruits. Second, research and experimentation should be encouraged on a tier of so-called minor crops and wild plants in various stages of domestication, including fish-bait trees, exotic fruits for the juice and vitamin trade, and timber trees. Many floodplain farmers have already taken the initiative planting the "next generation" of perennial cash crops, and their experience is worth tapping.

Among the crops with sure markets in Amazonia are cupuaçu, banana, and mango, all found in home gardens. Cupuaçu tolerates moderate flooding, and thrives on the higher parts of the floodplain. Banana and plantains are well adapted to floods and fruit throughout the year. The most important perennial crop along the middle Amazon, banana is usually planted in monocrop stands, but is also occasionally grown with other crops, such as cacao, manioc, maize, or papaya. I noted four banana cultivars—Prata (silver), Inajá (an upland palm), Branca (white), Casca Verde (green skin)—and one variety of plantain (Pacovão) in a two-hectare field intercropped with bitter manioc and tomato at Piracauera da Cima near Santarém. Banana succumbs to prolonged inundation, but soon recuperates as floodwaters recede and new shoots emerge. Floods assist the crop by destroying many of the pests and diseases, and by replenishing the soil.

Mango, originally from India, is widely grown in the Amazon and many small fruited seedling types have arisen, often with a high fiber content leaving the consumer with "stringy teeth." Grafted selections of mango with fewer dental floss-like strands and high market value, such as Haden and Keitt which were selected in backyards in Florida in the early 1900s, could be planted either in agroforestry plots or monocultural orchards on the Amazon floodplain. The relatively large-fruited Haden and Keitt varieties are already planted profitably in various states of Brazil, particularly in the Northeast. At present a number of local selections are cultivated in home gardens on the floodplain of the middle Amazon, such as Azul (blue), Bolocha (cracker), Breu (resin), Grande (big), Mangarita, Marquesa, (Marquise), Periquita (parakeet), and Roxinha (little red one), but their market value is lower than cultivars with larger fruits typical of the Florida varieties. Local selections are fine for home gardens, and agricultural development should not discourage their cultivation as a matter of policy. But for commercial use, farmers will want high quality planting material of varieties with strong market demand. Other popular fruit trees found in home gardens that could be planted on a larger scale, depending on local market conditions, include açaí, cashew, coconut, and guava.

A parallel effort should be undertaken to assess the potential of less well-known crops and plants currently in various stages of domestication in home gardens. Several unusual fruit trees are cultivated in home gardens or are gathered in the wild, and some of these could be developed commercially. But the process of selecting promising genotypes and securing a market for their products is time consuming and costly. Partnerships are needed between companies and scientific institutions to support the protracted research effort required to bring novel fruits to market.

Fish-bait trees are one promising avenue to pursue. Fish farming is a growing business in many parts of the Amazon, such as around Santarém. All of the fish farms are on uplands, not on floodplains where the fish normally occur. Operators at various scales are entering the business, ranging from small-scale farmers to ranchers. Attempts could be made to plant such fish-bait trees as araçá, baianeiro, catauarí (figure 6.7), curumi, socoró, tarumã, and uruá, around the

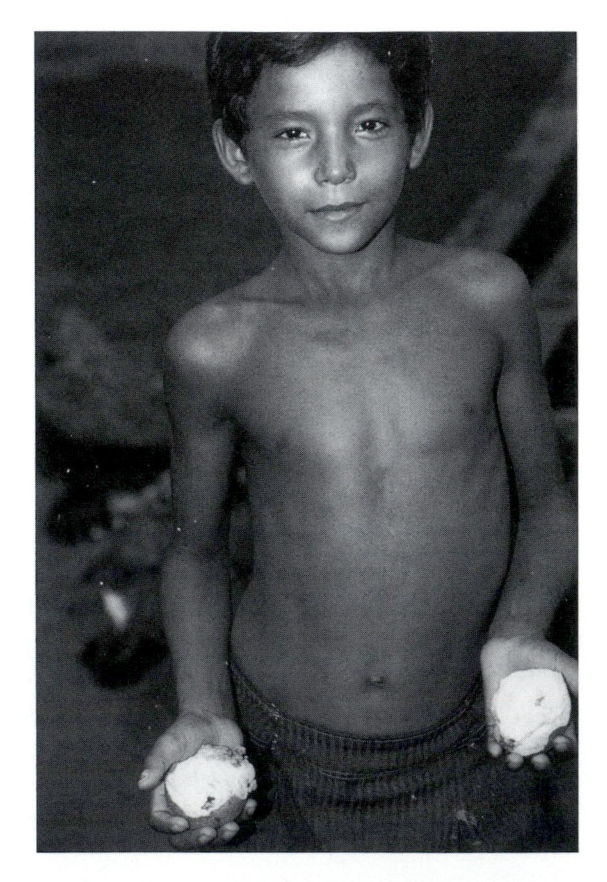

Figure 6.7. Catauarí fruit in a floodplain home garden. The catauarí tree arose spontaneously in the backyard and was protected because the fruits are used to catch tambaqui, one of the region's most prized fishes. Paraná do Baixo near Alenquer, Pará, June 1994.

Figure 6.8. A pond-raised tambaqui in a pioneering colonization zone of Rondônia. The farmer is holding a specimen kept for breeding purposes, while an agricultural extension agent on the right offers advice. The banks surrounding the ponds are bare; they could be planted to fruit trees to enrich the diet of the fish. Sitio Nossa Senhora Aparecida near Ji-Paraná, September 1993.

ponds. If the fish-bait species are not adapted to such environments, generic equivalents from upland forests might prove suitable.

One of the more popular pond raised fish is tambaqui, an omnivore in which fruits constitute an important part of the diet in the wild. At present, tambaqui (figure 6.8) are fed a variety of foods, including commercial fish feed, maize, manioc, and even snack foods with expired "sell-by" dates. In Ji-Paraná, Rondônia, pond-raised tambaqui are sold when they are relatively small, in the one- to two-kilogram stage, but consumers sometimes remark on their "muddy" flavor. River-caught tambaqui from the Guaporé are worth about 25 percent more in town because of their savory, nut-like flavor.

Eventually, however, the market for artificially raised tambaqui is likely to evolve to the larger specimens that yield "ribs." Tambaqui's delicate flesh is attached to sizable ribs that are especially popular with the well to do in cities for barbecuing. Tambaqui require at least 4 years to reach the "rib" stage, and such fish could at least be finished with a diet of fruit before going to market. Already, one small farmer near Ji-Paraná gathers seed from a small orchard of rubber trees to feed tambaqui stocked in his dammed stream. If it were not for the fish, he would probably have cut down the rubber trees because there is no money to be made from tapping the latex. And a rancher in Santarém plans to plant fruit trees, including açaí, around a tambaqui pond on his upland ranch. Floodplain growers could cultivate fish-bait trees to supply fruits to upland fish farms.

Camu-camu is another promising agroforestry candidate for floodplain farmers. A recently domesticated bush native to certain floodplain areas of central and western Amazonia, camu-camu produces bright-red, cherry-like fruits. Camu-camu is one of the region's up-and-coming fruits since it makes an enticing juice and is exceptionally rich in vitamin C. At the moment, though, upland growers in Amazonia and other parts of Brazil seem more likely to cash in on the market opportunities for this crop. In 1996, Fink established a nursery with a million camu-camu seedlings at km 100 of the Manaus-Itacoatiara highway. The Manaus-based company has contracted with about thirty nearby farmers to grow the crop. But most of the seedlings are being dispatched in plastic bags by overnight mail to São Paulo, where a 500-hectare plantation of the crop is being established with drip irrigation. Fink chose São Paulo as the mainstay of the operation because the soils are inherently more fertile than around Manaus and yields are expected to be higher. Bank loans are also much easier to obtain there than in Manaus, and São Paulo commands a huge urban market—with over 15 billion residents—and a growing demand for "natural" vitamin C pills. The paucity of infrastructure in Amazonia puts growers at a disadvantage compared to some other regions. In spite of these difficulties, innovative farmers and pioneering agroindustries are beginning to "jump-start" commercial agroforestry in Amazonia.

Another vocation of the floodplain is timber production, which could be feasible in agroforestry plots owned by small farmers, on larger plantations, or in forests owned by individuals or communities. In upland areas, some farmers are incorporating several timber species, such as mahogany and cedar, in agroforestry plots, particularly along the Transamazon Highway and in Rondônia. Timber trees are particularly noticeable in cacao and coffee groves in pioneering settlement areas. Several commercially valuable timber species are also adapted to the various flooding regimes of the Amazon floodplain and could be planted as medium and long term investments. With the growth of plywood factories in the Brazilian Amazon, the market for "soft" woods is expanding, creating more opportunities for farmers. Quick-growing trees, such as kapok, could be grown in agroforestry plots or in monocultural stands adjacent to other cash and subsistence crops. Virola, which is used for veneer, and mulato wood are particularly promising for hardwood markets because they are already well-known in the timber trade, especially virola which is logged heavily the estuarine area of the Amazon. Cajurana, another valuable timber tree of the floodplain, has the added advantage of producing fruits much sought after by commercially important fish, such as tambaqui.

As is frequently the case with "new" ideas for agricultural development, growers are often out in front of researchers and planners. A farmer at Piracauera da Cima near Santarém, for example, has encouraged spontaneous seedlings of mulato wood in an area devoted to vegetable production during the low-water season (figure 6.9). Eventually, he plans to harvest the wood for the timber trade. The seedlings arose spontaneously because the farmer has deliberately spared several hectares of forest on prime agricultural land. His woodlot also contains tachi trees (unidentified), the flowers of which attract honey-producing jandaíra bees. This farmer, as well as an appreciable number of other floodplain dwellers (figure 6.10), raise native, stingless bees in sections of hollow logs suspended from the rafters in their homes to garner the honey.

Figure 6.9. The spontaneous seedlings of mulato wood in the foreground are being encouraged in a floodplain vegetable plot because they will eventually produce valuable timber. The seeds dispersed from a nearby woodlot maintained by the farmer. Piracauera de Cima, near Santarém, Pará, July 1996.

Figure 6.10. A colony of stingless bees, known locally as *jupará*, raised for honey in a hollow log suspended from the ceiling of the kitchen area of a floodplain home. Although floodwaters have been receding for about a month, residents still have to step into canoes when they leave home. Ilha do Carmo, Municipality of Alenquer, Pará, July 1996.

Little if any thought has been given to the rational management of floodplain forests for timber and other products, nor to reforestation along the Amazon. Stocks of virola will likely be logged out of the lower Amazon within a few decades. Now is the time to conserve, manage, and plant timber resources, before they disappear to the detriment of floodplain residents.

Agroforestry could also embrace livestock production, rather than being seen as "opposing" land uses. In many parts of the tropics, tree crops and pasture are combined, but such instances are rare in Amazonia. Most of the cases are confined to the uplands, where some farmers graze cattle under the partial shade of tree crops, particularly rubber. The only example of agroforestry combined with grazing I witnessed along the Amazon was by the eastern shores of Marajó Island. Near Pesqueiro Beach about five kilometers north of Souré, some small farmers raise sheep under coconut palms planted on the floodplain of a creek. At full tide, the sheep have to retreat to higher ground. Given the importance of livestock to the economy of the Amazon River, scope exists for devising new and innovative ways to merge perennial crop production and the raising of animals.

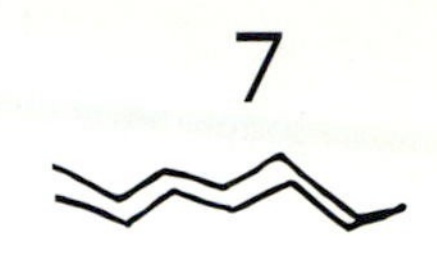

Restoring a Threatened Paradise

Successful economic development along the Amazon floodplain hinges on efforts to conserve and better manage the river's unique biodiversity. At present, much of the biodiversity is being destroyed even before it can be identified, let alone studied and evaluated for the benefit of those living in the region. The sinuous forests of the floodplain are a cornucopia for food, livestock feed, medicines, building materials, fish bait, and a host of other uses. In addition, the forests are a largely untapped reservoir of plants for agricultural improvement. Seasonally flooded forests are home to several economically important fruit trees, such açaí, buriti, and yellow mombim. Future attempts to raise yields and to improve the quality of these trees will require extensive tapping of genetic resources. The loss of wild populations of these long-appreciated fruit, nut, and timber trees on the floodplain not only undercuts the income-generating capacity and nutrition of locals, but also forecloses options for crop improvement.

An equally serious issue is the loss of indigenous knowledge of plant and animal resources. Although most aboriginal groups have long since disappeared along the Brazilian Amazon, small farmers and ranchers have knowledge of natural history that can help unlock the potential of new crops and pest control measures, among other attributes. Locals should be regarded as partners in managing and conserving biodiversity, rather than as obstacles to preserving it.

A wide array of issues surrounds the conservation and management of biodiversity along the entire spectrum from little-disturbed habitats to those heavily modified by human activity. This chapter explores some policy dimensions to biodiversity conservation and agricultural development. I begin with a discussion of habitat conservation, especially the roles of the private sector and community resource management. Second, I discuss in situ and ex situ approaches to conservation of plant resources, emphasizing the importance of farmer participation in such projects. Third, I summarize some of the characteristics of appropriate agricultural intensification. Next, I explore the relevance of property rights to farmers' ability to qualify for credit. Finally, I discuss the potential future of land use along the Amazon in the new millennium.

Habitat Conservation

Too often, those debating conservation separate into two camps: those "purists" who want to conserve natural habitats, and others who argue that conservation is a modern form of colonialism and of little relevance to developing regions. Debates then slide into a quagmire because participants fail to recognize that no sharp dividing lines exist between "natural" and "cultural" habitats. Much of the Amazon floodplain has been modified by human activity for thousands of years, but that does not negate the need to better manage biodiversity. And in those areas that are still relatively intact, conservation is especially important because biodiversity levels are likely to be high.

Some environmentalists may not be pleased with the contention that Amazonia was once densely settled, especially along the Amazon and other white-water rivers, and that most of the habitats in the basin today have felt the hand of man, even if they look "pristine." Opposition to this notion can be traced to two sources. First, if the Amazon is capable of supporting dense populations, rampant development could be given a green light. Some conservationists would likely feel more comfortable in the company of environmental determinists who argue that "sustainable" agriculture in the Amazon is largely futile, and that the few surviving unacculturated groups accurately represent pre-Columbian settlement patterns. According to that view, the region's ecosystems are too fragile, and the natural resource base too limited, to support dense human populations. The region is thus best left to the birds and the bees. Second, many ecological studies in Amazonia have been undertaken in environments that have been altered to varying degrees by people and some of those studying such topics as nutrient cycling or forest biomass have doubtless thought they were working in "natural" settings. In other words, many ecologists are reluctant to accept the notion that people have been rearranging the biological furniture in the Amazon for a long time, even in areas now covered by forest.

On the other hand, many agricultural scientists, planners, and nongovernment organizations (NGOs) concerned with social issues would likely be attracted to the assertion that the natural resource base of the Amazon is quite capable of supporting large and healthy human populations. Agricultural scientists might see another frontier to bring under the plow, while government agencies and NGOs that promote "community-based" development might argue that much of the region is a cultural forest and that conventional protected areas are meaningless. A great deal of potential exists for developing parts of the Amazon floodplain for agriculture. The issue is how it should be done, given the environmental and social costs.

The proposition that Amazonia should be abandoned to laissez-faire occupation, unfettered by parks and reserves, is untenable. Although little of the region can be considered pristine, this fact should not be seen as a carte blanche to enter and further alter all of the region's diverse habitats. Some parts of the Amazon have been little altered for centuries, especially after the collapse of indigenous populations following contact with Europeans. Such mature habitats warrant protection. Conservation in Amazonia should include diverse environments with varying degrees of human disturbance. The notion that protected areas are an elitist imposition is unwarranted. Some areas deserve protection from clearing, hunting, and harvesting of forest products. Other conservation units would provide for various extractive activities.

Different segments of the Amazon River need to be preserved to serve as food and shelter for many fish species, as genetic reservoirs, and as sources of numerous products for locals. Two main types of reserves are needed: nature preserves with little or no human disturbance and multiple-use reserves where certain animal and plant products are harvested without seriously affecting the integrity of the ecosystems.

The marked ecological heterogeneity of the floodplain calls for a multilocation approach to conservation, but only three reserves have been established along the Amazon. The Mamirauá ecological station near Tefé, created in 1990, covers 1.2 million hectares and is the world's largest freshwater reserve. Preexisting residents within the reserve are considered legitimate resource managers rather than agents of ecological destruction to be expelled. Only time will tell whether residents will be able to keep out commercial fishing fleets and loggers, and whether they will "manage" natural resources within the reserve without undercutting their ecological viability. Mamirauá nevertheless stands as a high-profile political statement: It is important to safeguard habitats along the floodplain, and on a grand scale.

Adjacent to Mamirauá, is the Amana reserve, created in 1998 by the Amazonas state government. The Amana reserve straddles parts of the Amazon floodplain near the confluence with the Japurá and extends inland to the Jaú reserve in the Rio Negro basin. The Amana reserve, which like Mamirauá calls for preexisting residents to participate in the management of the reserve, covers 2.3 million hectares. The Amana, Mamirauá, and Jaú reserves serve as a broad ecological corridor embracing 22,000 square kilometers—the largest contiguous protected forest on earth.

The third protected area, the Pacaya Samiria reserve, is located in the wedge between the confluence of the Ucayali and Marañon rivers in Peru. Established in 1982, the reserve covers 20,800 square kilometers, 80 percent of which is floodplain. As in Mamirauá and Amana, the Pacaya Samiria reserve embraces preexisting settlers, composed mostly of the Cocama indigenous group. Some 33,000 people, distributed among one hundred communities, farm and fish in the reserve. Outsiders, especially commercial fishing fleets, regularly penetrate the poorly patrolled reserve and it remains to be seen whether communities, supported by external funding and technical advice, will be able to safeguard the reserve's natural resources.

Many other biologically interesting areas of the Amazon floodplain lack any formal protection or efforts to promote sustainable harvesting of natural resources. Much work remains to be done to identify high-priority areas for parks and reserves. But which areas are feasible as protected or managed zones will depend on input from a variety stakeholders, from local communities to ranchers and loggers.

Reliance on government-designated parks or reserves as the sole approach to conserving biodiversity on the floodplain is a risky strategy. Political support for reserves is often shaky, especially at the local and state levels. When austerity measures are announced, conservation programs, education, and research are typically among the hardest hit. Among local politicians and business owners a perception has emerged that the federal government is "pushing" parks and reserves in response to "green" pressure from abroad. Conservation of natural resources and habitat preservation is not a high priority for most of them. Also, farmers, fishermen, and loggers typically see parks and reserves as an open invitation for resource exploitation.

Role of the Private Sector

Although further work in identifying areas on the Amazon floodplain for parks and reserves is warranted, major emphasis should be placed on working with the private sector to achieve the goal of conserving and better utilizing the region's biodiversity. The private sector involves all stakeholders on the Amazon floodplain, from small-scale growers to corporate farmers and ranchers.

Large landowners can play a dual role in conservation. As powerbrokers, their support for the creation of parks and reserves is needed. And perhaps more importantly, they can become involved directly in helping to conserve biodiversity because their properties often contain sizable patches of relatively undisturbed forest and other habitats. Along the Manaus to Monte Alegre stretch of the Amazon, for example, many of the sizable patches of floodplain forest are in the hands of ranchers. Ironically, while ranchers are the major agents of forest destruction, they are also one of the keys to its survival.

Forest patches on ranches are particularly valuable from the point of view of conservation for two reasons. First, they are relatively large, which is especially important for plants and animals that require generous areas of intact habitat for survival. The greater the degree of habitat fragmentation, the greater the loss of species, due to the disappearance of pollinators and seed dispersal agents, among other factors. Second, the best-preserved patches of forest tend to be on the higher banks, areas that are typically cleared by farmers. Plant and animal communities change along a gradient from high banks down a gentle slope toward the backswamps. Ranches thus encompass most of the remaining mature forest adapted to relatively brief periods of inundation.

In areas of the floodplain dominated by small farmers, sizable stretches of forest can sometimes still be found, but usually in backswamps. On higher banks the forest has largely been cleared for houses, home gardens, fields, and pasture. While adjacent properties can encompass a relatively large belt of backswamp forest, individuals may decide to clear the entire property, breaking up the forest into a checkerboard pattern of islands. Ranchers, on the other hand, control the destiny of the contiguous forest on their property.

Both ranchers and individual farmers can "protect" the forest on their holdings. Some ranchers take pride in the forest and lakes on their properties, and take appropriate measures to prevent logging and other unauthorized activities. If a community is well organized, it too may be able to keep out intruders. But both parties can be tempted to cut forest to make money.

Efforts to promote conservation on the floodplain should adopt the carrot approach rather than the stick. A command-and-control effort from central governments to enforce environmental legislation has been largely ineffective, at least in rural areas of Amazonia. Rather than punish individuals and corporations for cutting down forest on their properties, incentives should be offered to encourage them to maintain forests. For this approach to work effectively with large landholders, however, Brazil's tax collection system will need to be revamped. People will only be motivated to save forest if it pays, such as by a significant reduction in their tax bills. Many of the larger entrepreneurs on the Amazon floodplain find ways to keep their tax bills negligible anyway. And small farmers generally do not pay any income taxes.

Perhaps a short-term solution to this problem would be to provide direct payments to certain ranchers, communities, and individual farmers for maintaining forests on their land. While there would still be scope for corruption, deforestation could be monitored by satellites. This practice is already used to identify landowners who are cutting and burning forest illegally. Why not use the same technology to simply reward those who keep land in forest? Conversely, if farmers, communities, and ranchers wish to clear all of their properties, they should be free to do so. The limited budgets available for environmental conservation should be directed towards providing incentives for safeguarding habitats rather than extracting fines.

Community Resource Management

At present much of rural development in Amazonia focuses on "communities"; community "resource management" has become a veritable rallying cry for government policymakers, multilateral development banks, bilateral aid agencies, and NGOs. Community-based resource management is being pursued in Amazonia, and in many other tropical regions, with almost missionary zeal. But progress in conservation and development in Amazonia, as elsewhere, will only come with an open and questioning mind, not blind faith. For example, environmental change in Amazonia is being wrought as much by business people residing in urban centers as it is by peasants in the countryside.

Community is an imprecise term, like *sustainable development*. Most of the villages and farming areas I have visited in the Amazon contain a range of actors with different income levels and agendas. In the past, at least, the stampede to promote community management of natural resources has tended to assume that communities are "anti-capitalist," when in fact there is little evidence for this in the Brazilian Amazon. On the contrary, I have yet to meet a family of farmers or fisherfolk not interested in making money. One cannot even count on religious homogeneity in rural populations to help encourage community cohesiveness; many nominal Catholics have converted to various Protestant denominations, especially Pentecostal groups. Belatedly, some proponents of community resource management are recognizing the heterogeneity of their "clients," but then only too often the issue becomes one of identifying the "right" group or groups, and working exclusively with them.

Certainly, communities that seek outside assistance to manage their natural resources deserve support, including training. A mosaic of approaches to conservation and development in Amazon is needed, ranging from community-based management systems where appropriate, to individual smallholders and larger landowners. Yet, a rigorous analysis is needed of experiences with community management and protection of natural resources in the Amazon so that the pitfalls and potential of such approaches can be better understood.

Agrobiodiversity and Crop Genetic Resources: A Conservation Imperative

Biodiversity conservation also applies to crop production. Yet some people argue that agricultural habitats are so transformed that biodiversity is no longer an issue; instead, agriculture must be "contained" to minimize its impact on surrounding habitats. But cropping systems along the Amazon floodplain are diverse, contain a rich array of species, and in some cases, numerous varieties.

Two approaches to preserving agrobiodiversity are typically found: ex situ gene banks, in which plant materials are stored away from areas where they are grown; and in situ conservation, in which plants are preserved in farmers' fields and home gardens. Both approaches are warranted, but in the case of the Amazon floodplain, in situ conservation should receive the highest priority. Gene-bank conservation works best for crops with seeds that can be easily dried and frozen for long-term storage. But which seeds to store? The main seed crops of the floodplain, rice and maize, are not indigenous to the area. It is possible that farmers have selected locally adapted varieties of maize, but no surveys have focused on that issue along the Amazon. It seems likely that some of the squashes growing along the middle Amazon are unique and warrant collection for seed gene banks. But for the most part, in situ approaches are warranted because many of the important crops in the region are propagated by cuttings or are perennials that produce seeds that cannot be dried and frozen without killing them.

The basic staple along the Amazon, manioc, can be kept as living plants in field gene banks at research stations, but such collections are costly to maintain because the material has to be kept in open fields or in tissue cultures. Few of the manioc cultivars encountered in the Amazon are in field gene banks at agricultural research stations. Of the twenty-two sweet manioc cultivars I encountered in the Amazon, only about eight appear to be in the manioc collection at CIAT (Centro Internacional de Agricultura Tropical) in Cali, Colombia, the world's largest assemblage of manioc germ plasm. Even those cultivars with matching names, such as Amarelinha and Branca, are common in other parts of Brazil and may be genetically distinct from those I encountered in the Amazon. Of the 101 cultivar names for bitter and sweet manioc that I noted in the field, only fifteen of the names are matched among 269 accessions held in the field gene bank for the crop maintained by EMBRAPA, Brazil's agricultural research service, in Belém. And those fifteen matches may not be identical genotypes.

The task of sorting out the numerous manioc cultivars in Amazonia, and identifying synonyms for them, has barely begun. CIAT's collection emphasizes sweet manioc, whereas bitter types are economically more important in the Brazilian Amazon. Financing crop germ-plasm conservation is often precarious in agricultural research programs, especially at national and state institutions. Funding shortfalls often lead to the loss of accessions in gene banks.

EMBRAPA's center in Belém is describing and evaluating the manioc collection in its care. Such laudable efforts need to be augmented because interesting material can come to light. For example, EMBRAPA has chosen Flor de Boi, a favored bitter manioc among floodplain farmers in the Santarém and Monte Alegre areas of the Amazon, for breeding work. Flor de Boi is precocious and high yielding, producing twenty-four tons of tubers per hectare within just six months.

Cultural information about gene-bank accessions is almost always lacking. For most accessions only "passport" data are available, that is, the location where the sample was collected and cursory information on vegetation and soils. How the varieties fit among the constellation of other cultivars, and their role in dietary customs and religious practices, is usually missing. The cultural attributes of varieties also warrant study, even though social scientists are rarely included on germ plasm–collecting trips.

For the most part, agricultural intensification along the floodplain involves nonnative crops, including vegetables, and genetic erosion is not especially marked. Eventually, however, farmers will begin to adopt practices that streamline not only the number of crops grown, but the varieties as well. The problem is that many of those crops—especially the ninety or so perennials found in home gardens—cannot be conserved in conventional seed gene banks. Much remains to be done to better understand how to mitigate biodiversity losses as farmers intensify their operations.

Agricultural Intensification

The Amazon floodplain can be developed for increased agricultural production while still maintaining a diverse array of habitats. This was accomplished long before the arrival of Europeans in the early sixteenth century, and can be attained again with a better appreciation for cultural systems and natural history along the Amazon.

Lessons from Traditional Food-Production Systems

Experience has shown that intensifying food production in Amazonia with heavy reliance on machinery and purchased chemicals for crop protection is risky. The low prices that basic staples fetch in the market endanger any farming with crops that requires heavy investment. Much can be learned about how traditional food-producing systems are changing on the floodplain in response to the growing market for staples in towns and cities. Some of the low-cost strategies employed by small farmers could probably be modified to increase yields without damaging the environment.

Development policies for food production in Amazonia should nevertheless avoid romantic ideas about returning to traditional methods or just promoting existing, low-input systems. Static notions about "premodern" agriculture fail to consider the dynamism of all agricultural systems, including so-called traditional ones. Rather, a fuller appreciation of the range of technologies and management strategies of traditional systems is needed as a precursor to any interventions. The marketing and price structures for foodstuffs in the Amazon also need checking to determine whether artificially low food prices discourage farmers from producing staples for the market.

Open Systems Are More Resilient

Although many of the keys to a more rational development of the Amazon floodplain can be found in the native plants and animals, it would be unwise to close the door to introducing crops, varieties, and livestock breeds from other regions. Resilient systems tend to be open to incorporating novel information. Already, the mix of cultivated plants and livestock on the floodplain is derived from both indigenous and exotic species, and the process of assimilating new and promising plants and animals from other lands will continue. In the Bolivian Amazon, the Brazilian geographer Mário Hiraoka has found that the success of farmers hinges on their the willingness to adopt new crops and farm management practices, and the same applies for most regions of the world.[1]

In parts of Thailand, for example, farmers have selected rice varieties that tolerate rapidly rising waters, and some of these deepwater rices might be appropriate for the Amazon floodplain. Some Thai rice tolerates submergence for extended periods as it "catches up" with floods, so it could be grown in areas along the Amazon not normally cultivated, thereby leaving the higher parts of the floodplain in forest, or if already cleared, to other crops. Around Lake Chad in West Africa, pastoralists have selected a specialized breed of cattle, the Kuri, that can even feed under water. The Kuri breed has exceptionally large, buoyant horns that help the animals float as they seek out submerged forage. Cattle production on the Amazon floodplain could thus be intensified without the need to clear more forest.

Farmers as Participants in Research and Development

Although agroforestry is one of the more promising avenues available for promoting environmentally sound development along the Amazon, development planners and executing agencies, including government bodies and NGOs, should avoid trying to pick "winners" for such systems. Farmers are more in tune with economic realities than agricultural research and extension agencies and are in a better position to select the components for their agroforestry systems. Governments can help by eliminating biases in the fiscal and regulatory environment against tree planting, or can at least even the playing field vis-à-vis other land-use systems.

The tendency at agricultural research centers is to design a single agroforestry model, then to study the system in replicated trials. Even when farmers are involved in the process, they are typically passive observers rather than active participants. While data accumulate for annual reports, the systems themselves often prove to be irrelevant to realities confronted by farmers as markets and other environmental conditions are constantly changing.

Agricultural research systems need to become more agile and responsive to farmers to backstop crops currently in production, and to provide good planting material for the next generation of crops and varieties. At present, much of the technology deployed by farmers along the Amazon is coming from the informal or private sector. If farmers are involved at the outset in the design of agricultural research projects, these projects are more likely to meet their needs. And the farmers themselves can serve as extension agents for promoting sound practices.

Agrobiodiversity Surveys

One way to devise agricultural development projects that help conserve the natural resource base while addressing the needs of local farmers is to conduct agrobiodiversity surveys during the project's initial stages. One of the merits of such an undertaking is that it provides data from which to quantify trends. A range of disciplines—from genetics, botany, agronomy, ecology, and the social sciences—need to be tapped to understand the dynamics of genetic erosion and to devise appropriate strategies to manage agrobiodiversity. The composition of agrobiodiversity survey teams would vary, but at a minimum they should contain an economist, an ethnobotanist or ethnoecologist, an ecologist, a social scientist, and some local farmers and livestock owners. Agrobiodiversity teams would assess the diversity of breeds and varieties; identify habitats that provide environmental services, such as watershed protection and havens for crop pollinators; ascertain whether any habitats serve as reservoirs for wild populations of crops and their near relatives; and document the degree to which locals gather products from uncultivated areas.

An agrobiodiversity survey should not be perceived as a Noah's ark approach to genetic resources, designed to rescue materials about to be swamped by new and improved technologies. Rather, the idea is to understand the existing patterns of agrobiodiversity management and the role that diverse combinations of plants and animals play in sustaining the livelihoods of locals. It should be possible, then, to weigh the comparative advantages of traditional and improved crop varieties and livestock breeds, and to alert planners to the dangers of losing unique germ plasm. If enough agrobiodiversity surveys are undertaken, and the findings are entered into computer databases and GIS (Geographic Information Systems) software, it should be possible to cross-reference varieties and crops that might be transferable to areas with similar soils and climate.

Property Rights and Credit

To promote agroforestry on the Amazon floodplain, several steps need to be taken in addition to conserving agrobiodiversity. Promising crops need to be multiplied in well-run nurseries and distributed to farmers. Specially targeted credit programs are needed to help farmers adopt commercial agroforestry and monocultural orchards, and to help entrepreneurs establish agroindustries to process fruits and prepare nuts and other products for regional, national, and international markets.

In order to obtain credit, a person must have title to the land. Most small farmers appear to have no documentation for the land they occupy. Land titles, in fact, are often vague and confusing because a plethora of documents with varying degrees of authority have been issued for land over the centuries. At the turn of the century, for example, the state of Pará was offering at least two types of land title: *Título de Posse* (occupancy title) and *Títutlo de Legitimação* (legitimate title). The first document recognized that a person had undisputed claim to the land. After some six years or so, if no other claimants had surfaced, the owner could press claims for a more definitive second title with the Secretary of Public Works, Land, and Transportation (Secretaria de Estado de Obras Públicas, Terras e Viação). Yet the same piece of land may have several claimants because it has been—and still is—relatively easy to file papers with county government offices (*cartórios*). The state land-title agency in Pará is now called ITERPA (Instituto de Terras do Estado do Pará); and in addition to ITERPA documents, some landowners have obtained titles from the federal land agency, INCRA (Instituto Nacional de Colonização e Reforma Agrária).

In some cases, such as in Santana do Ituqui, parts of the *várzea* are farmed individually, although the land "belongs" to the community. Newcomers request permission to till sections of Ituqui Island opposite the town. With increasing interest in developing the *várzea*, the ill-defined nature of land titles may lead to problems in the future and could serve as a disincentive for small farmers to invest in perennial cropping systems.

Although the land tenure picture remains confused in many parts of the Brazilian Amazon, the floodplain is not the scene of violent clashes over land rights. Government policy has steered migrants to the uplands of Amazonia, particularly southern Pará, Rondônia, and Acre, where conflicts over land have arisen. On the floodplain, investors are buying out small farmers and ranches, rather than acquiring dubious land "titles" and then expelling the occupants. Investors buy out the small farmers even if they lack land titles.

Demarcation of land on the floodplain is a challenging task. The restless *várzea* changes its configuration constantly, and old maps can quickly become useless for navigation. Some islands disappear, while others are born. One Santarém citizen complained that he is still paying taxes on a small piece of property that he inherited from his father, even though the Amazon long ago eroded his small ranch. Tax officials refuse to recognize this natural hazard of floodplain life, and the perplexed citizen continues to fork out taxes on the deceased ranch for fear that a lien will be placed on his other properties.

Rather than call for land reform, the answer offered by many development debates, governments need to put mechanisms in place to ensure resource rights rather than the promulgation of official land titles. In most areas on the middle Amazon floodplain, ownership rights are recognized peacefully, even though many of the farmers and ranchers have dubious documents or none at all. The most urgent need is to provide financing for people whose land claims are respected locally, rather than wait interminably for some government agency to conduct property surveys.

More support is warranted for agroindustries to process floodplain tree crops such as frozen fruit juices, fruit-flavored yogurts, and essential oils, since little credit is available for microenterprises in the Brazilian Amazon. Some processing plants could be assembled on barges so that they can be towed to areas of high demand for their services. If people are making money growing a wide variety of fruits and nuts on the floodplain, they are more likely to be motivated to conserve such resources, including forests.

The Floodplain in the New Millennium

Landscapes along the Amazon have changed dramatically over the millennia, and they will continue to do so. In the short term, floodplain forests will continue to shrink as cattle and water buffalo ranching expand. Eventually, though, the importance of large livestock on the floodplain will diminish due to dropping per capita consumption of beef and other more attractive options for ranchers. Both perennial and annual cropping will expand, at the expense of pasture rather than forest.

As world demand for hardwoods outstrips supplies, more entrepreneurs will invest in timber plantations on both the floodplain and uplands. Small and large operators alike will plant a diverse array of trees for the plywood and veneer markets. Some of the operations are likely to be large single-species plantations, while more farmers are likely to incorporate a number of timber species in their agroforestry fields.

Fish farming will increase as fisheries are depleted and as urban consumers seek healthier diets. Cultivation of fish-bait fruits, harvested from agroforestry plots and single-species orchards, will then become a profitable enterprise. Fish farming will likely become widespread on the floodplain, in large cages or other enclosures. The environment can only provide "free rent" up to a point. Even if fish management becomes an art along the Amazon, there will still be a market for raising high-value fish such as tambaqui in captivity.

Although perennial cropping for timber and other products, especially fruits, will likely increase, the cultivated area will still be dominated by annual crops for some time to come. Mechanized soybean production will likely emerge as a major economic activity in the early part of the twenty-first century. Thirty years ago, no soybeans were grown in the Brazilian Amazon outside of a handful of research stations. Now soybean is big business on the uplands in southern Rondônia, southern Pará, and along the Carajás-Itaqui railroad in Maranhão. The soybean front is moving north, and will eventually prosper on the fertile soils of the floodplain. Rice production will most likely expand also, particularly on large mechanized farms. The number of small farms will diminish, as it has in most other parts of the world where market influences are strong. Small farmers will have a comparative advantage in agroforestry and manioc production, which are labor intensive.

The above scenario, with its diverse mix of crop and animal husbandry, would be completely scrambled if the Amazon was ever dammed. As far-fetched as that idea might seem, it has on occasion been contemplated. Herman Kahn of the

Hudson Institute in New York proposed such an idea in the 1960s; according to that vision, a huge lake would be formed to facilitate navigation and gain better access to upland resources.[2] The idea died, not because of its ecological lunacy, but because it was proposed by a gringo and triggered concerns about sovereignty.

Although no such dam is contemplated at the moment, such a scheme could resurface. Within the last few decades, several tributaries have been dammed to generate hydroelectricity, such as the Tocantins, Curuá-Una, and Uatumã near Manaus. Politicians find megaprojects attractive, and so do contractors. Other continental-sized rivers, such as the Nile, have been dammed in the last half century, and the silt-laden Yangtze in China is about to suffer the same fate. The Three Gorges Dam on the Yangtze is projected to cost around $25 billion, even though the huge reservoir might fill with silt in as little as ten years. The Three Gorges Dam has been touted as a triumph of socialism; an Amazon dam could just as easily be heralded as a capitalist windfall.

It is to be hoped that the reemergence of democracy in Brazil will thwart any attempts to implement a dam-building project along the Amazon. The United States initially spent billions of dollars channeling the Mississippi through artificial levees. Once in a while, the river bursts through its straitjacket, causing millions of dollars of flood damage, as happened in 1993. Millions of dollars are then spent shoring up the dikes until the next time Nature reasserts itself. Let us hope that cultures along the Amazon continue to evolve land-use adaptations to the annual floods, rather than try to eliminate them.

Appendix A

Common and Scientific Names of Plants

English	Brazilian	Scientific
Acapurana	Acapurana	*Campsiandra comosa*
Açaí palm	Açaí	*Euterpe oleracea*
Albina	Albina	*Turnera ulmifolia*
Amazon willow	Oeirana	*Salix martiana*
Ambarella	Cajarana	*Spondias dulcis*
American oil palm	Caiaué	*Elaeis oleifera*
Ananí	Ananí	*Symphonia globulifera*
Andirá-uxi	Andirá-uxi	*Andira retusa*
Andiroba	Andiroba	*Carapa guianensis*
Annatto	Urucu	*Bixa orellana*
Apé	Apé	*Urospatha caudata*
Apuí	Apuí	*Clusia* sp.
Arabica coffee	Café	*Coffea arabica*
Araçá	Araçá	*Myrcia fallax*
Aruanãzeiro	Aruanãzeiro	*Campnosperma* sp.
Arum	Aninga	*Montrichardia arborescens*
Assacu	Assacu	*Hura crepitans*
Assacu-rana	Assacu-rana	*Erythrina glauca*
Avocado	Abacate	*Persea americana*
Babaçu palm	Babaçu	*Attalea speciosa*
Bacaba palm	Bacaba	*Oenocarpus bacaba*
Bacabinha palm	Bacabinha	*Oenocarpus minor*
Bacuri	Bacuri	*Rheedia brasiliensis*
Baianeiro	Baianeiro, arapari	*Macrolobium acaciaefolium*
Barbados cherry	Acerola	*Malpighia glabra*

(*continued*)

English	Brazilian	Scientific
Biribá	Biribá	*Rollinia deliciosa*
Black-bellied tree duck grass	Capim marreca	*Paspalum conjugatum*
Boteiro	Boteiro	*Lonchocarpus denudatus*
Bow-wood	Pau d'arco	*Tabebuia serratifolia*
Brachiarão	Brachiarão, brizantão	*Brachiaria brizantha*
Brazil nut	Castanheira	*Bertholletia excelsa*
Breadfruit	Fruta-pão	*Artocarpus altilis*
Buriti palm	Miriti, buriti	*Mauritia flexousa*
Bussú palm	Bussú	*Manicaria saccifera*
Caçeira	Caçeira	*Cassia* sp.
Caimbé	Caimbé	*Sorocea duckei*
Caimbé-rana	Caimbé-rana	*Coussapoa asperifolia*
Calabash gourd	Cuieira	*Crescentia cujete*
Camapu	Camapu	*Physalis angulata*
Camu-camu	Camu-camu, sarão	*Myrciaria dubia*
Canarana grass	Canarana	*Echinochloa polystachya*
Canarana lisa grass	Canarana lisa	*Echinochloa pyramidalis*
Cannonball tree	Castanha de macaco	*Couroupita guianensis*
Capsicum pepper	Pimenta	*Capsicum frutescens*
Carambola	Carambola	*Averrhoa carambola*
Caraná palm	Caraná	*Mauritiella aculeata*
Carnaúba palm	Carnaúba palm	*Copernicia prunifera*
Cashew	Caju	*Anacardium occidentale*
Catauarí	Catauarí	*Crataeva benthami*
Caxinguba	Caxinguba	*Ficus anthelminthica*
Cecropia	Embaúba	*Cecropia latiloba*
Cedar	Cedro	*Cedrela odorata*
Cinzeiro	Cinzeiro	*Terminalia tanibouca*
Cipó ambé	Cipó ambé	*Philodendron imbe*
Coconut	Côco	*Cocos nucifera*
Cocoyam	Taioba	*Xanthosoma* sp.
Common bean	Feijão	*Phaseolus vulgaris*
Coriander	Coentro	*Coriandrum sativum*
Cotton	Algodão	*Gossypium* spp.
Cramuri	Cramuri	*Gymnoluma glabrescens*
Cumacaí	Cumacaí	*Lophostoma calophylloides*
Cupuaçu	Cupuaçu	*Theobroma grandiflorum*
Curuá palm	Curuá	*Attalea spectabilis*
Curumi	Curumi	*Muntingia calabura*
Cutitiribá	Cutitiribá	*Radlkoferella rivicoa*
Espinheiro	Espinheiro	*Acacia polyphylla*
Favarana	Favarana	*Crudia amazonica*
Flecheira	Flecheira	*Gynerium sagittatum*
Floodplain tortoise-tree	Jabutí da várzea	*Erisma calcaratum*
Genipap	Jenipapo	*Genipa americana*
Guava	Goiaba	*Psidium guajava*
Guarumã	Guarumã	*Ischnosiphon obliquus*
Guinea grass	Colonião, tobiatão	*Panicum maximum*
Inajá palm	Inajá	*Attalea maripa*
Ingá	Ingá	*Inga* sp.
Ingá baú	Ingá baú	*Inga cinnamomea*
Ingá capuchino	Ingá capuchino	*Inga laurina*
Ingá cipó	Ingá cipó	*Inga edulis*
Ingá grande	Ingá grande	*Inga* sp.

English	Brazilian	Scientific
Ingá xixi	Ingá xixi	*Inga heterophylla*
Itaúba	Itaúba	*Mezilaurus itauba*
Jacaréuba	Jacaréuba	*Calophyllum brasiliensis*
Jacitara palm	Jacitara	*Desmoncus polyacanthos*
Jambolan	Ameixa	*Syzygium cumini*
Jambu	Jambu	*Spilanthes oleracea*
Jauari palm	Jauari	*Astrocaryum jauari*
Jipioca	Jipioca	*Entada polyphylla*
Jucá	Jucá	*Caesalpinia ferrea* var. *Cearensis*
Junco	Junco	*Cyperus* sp.
Jupati palm	Jupati palm	*Raphia taedigera*
Jutaí	Jutaí	*Hymenaea* spp.
Jute	Juta	*Corchorus capsularis*
Kapok tree	Sumaúma	*Ceiba pentandra*
Lime	Limão	*Citrus aurantifolia*
Limorana	Limorana	*Chlorophora tinctoria*
Louro-inhamuí	Louro-inhamuí	*Ocotea cymbarum*
Maçaranduba	Maçaranduba	*Manilkara huberi*
Mahogany	Mogno	*Swietenia macrophylla*
Maize	Milho	*Zea mays*
Malay apple	Jambo	*Eugenia malaccensis*
Mango	Manga	*Mangifera indica*
Manioc	Mandioca, macaxeira, aypí	*Manihot esculenta*
Marajá palm	Marajá	*Bactris maraja*
Marimari	Marimari	*Cassia leiandra*
Matutí	Matutí	*Pterocarpus officinalis*
Munguba	Munguba	*Pseudobombax munguba*
Mulato wood	Pau mulato	*Calycophyllum spruceanum*
Murí grass	Murí	*Paspalum fasciculatum*
Muruci	Muruci	*Byrsonima crassifolia*
Murumuru palm	Murumuru	*Astrocaryum murumuru*
New World yam	Cará	*Dioscorea trifida*
Pancuã grass	Pancuã	*Paspalum furcatum*
Papaya	Mamão	*Carica papaya*
Pará grass	Colônia, mojuí	*Brachiaria mutica*
Paracaxi	Paracaxi	*Pentaclethra macroloba*
Paricá da várzea	Paricá da várzea	*Pithecellobium amazonicum*
Passionfruit	Maracujá	*Passiflora edulis*
Patazana	Patazana, pariri	*Thalia geniculata*
Pau de Angola	Pau de Angola	*Piper suffitor*
Peach palm	Pupunha	*Bactris gasipaes*
Pemembeca grass	Pemembeca	*Paspalum repens*
Pião branco	Pião branco	*Jatropha curcas*
Pião roxo	Pião roxo	*Jatropha gossypifolia*
Pineapple	Abacaxi	*Ananas comosus*
Piranha-tree	Piranheira	*Piranhea trifoliata*
Pitomba	Pitomba	*Talisia esculenta*
Pojó	Pojó	*Guazuma ulmifolia*
Pupunharana palm	Pupunharana	*Bactris macana*
Royal palm	Palma real	*Roystonea oleracea*
Rubber	Seringa	*Hevea brasiliensis*
Runner bean	Feijão metro	*Phaseolus coccineus*
Sandpaper tree	Caimbé	*Curatella americana*

(*continued*)

English	Brazilian	Scientific
Sapodilla	Sapotilha	*Manilkara achras*
Sapucaia	Sapucaia, castanha	*Lecythis pisonis*
Sapupira	Sapupira	*Diplotropis martiusii*
Socoró	Socoró	*Mouriri ulei*
Sororoca	Sororoca	*Ravenala guianensis*
Soursop	Graviola	*Annona muricata*
Squash	Abobra	*Cucurbita* sp.
Surinam cherry	Grosela	*Eugenia uniflora*
Sweet orange	Laranja	*Citrus sinensis*
Sweet potato	Batata doce	*Ipomoea batatas*
Sweetsop	Ata	*Annona squamosa*
Sugarcane	Cana	*Saccharum* sp.
Taboka	Taboka	*Guadua* sp.
Tamarind	Tamarindo	*Tamarindus indica*
Tangerine	Tangerina	*Citrus reticulata*
Tarumã	Tarumã	*Vitex cymosa*
Tucumã palm	Tucumã	*Astrocaryum vulgare*
Urena	Malva	*Urena lobata*
Uruá	Uruá	*Cordia nodosa*
Urucuri palm	Urucuri, arucuri	*Attalea phalerata*
Vassoura	Vassoura	*Scoparia dulcis*
Virola	Ucuúba	*Virola surinamensis*
Water melon	Melancia	*Citrulus lanatus*
Wild rice	Arroz bravo	*Oryza* spp.
Yellow mombim	Taperebá, cajá	*Spondias mombim*

Appendix B

Common and Scientific Names of Animals

English	Brazilian	Scientific
	Mammals	
Amazon bamboo rat	Saiuá, soyá	*Dacylomys dactylinus*
Black rat	Rato	*Rattus rattus*
Capybara	Capivara	*Hydrochaeris hydrochaeris*
Gray brocket deer	Fuboca	*Mazama gouazoubira*
Manatee	Peixe-boi	*Trichechus inunguis*
Paca	Paca	*Agouti paca*
Red brocket deer	Veado	*Mazama americana*
Tapir	Anta	*Tapirus terrestris*
	Birds	
Black-bellied tree duck	Marreca	*Dendrocygna autumnalis*
Black vulture	Urubu	*Coragyps atratus*
Blue and yellow macaw	Canindé	*Ara ararauna*
Brazilian duck	Pato	*Amazonetta brasiliensis*
Harpy eagle	Gavião real	*Harpia harpyja*
Helmeted guineafowl	Galinha d'Angola	*Numida meleagris*
Muscovy duck	Pato do mato	*Cairina moschata*
	Fish	
Aracu	Aracu	Species of *Leporinus, Rhytiodus*, and *Schizodon*
Jatuarana	Jatuarana, matrinchão	*Brycon* sp.

(*continued*)

English	Brazilian	Scientific
	Fish	
Pacu	Pacu	Species of *Metynnis* and *Mylossoma*
Pirapitinga	Pirapitinga	*Colossoma bidens*
Pirarucu	Pirarucu	*Arapaima gigas*
Sardinha	Sardinha	*Triportheus angulatus*
Tambaqui	Tambaqui	*Colossoma macropomum*
	Reptiles	
Cabeçudo	Cabeçudo	*Phractocephalus dumeriliana*
Giant river turtle	Tartaruga	*Podocnemis expansa*
Pitiú turtle	Pitiú, iaçá	*Podocnemis sextuberculata*
Yellow-spotted Amazon turtle	Tracajá	*Podocnemis unifilis*
	Insects	
Jandaíra bee	Jandaíra	*Melipona rufiventris*
Jupará bee	Jupará	*Melipona rufiventris* ?
Palm beetle	Suri (Peru)	*Rhynchophorus palmarum*

Appendix C

Seventy-nine Varieties of Bitter Manioc Cultivated in the Brazilian Amazon

Variety	Upland Location(s)	Amazon Floodplain Location(s)
Abacatinho		Ituqui near Santarém, Pará
Açaí	Near Rio Camará, Marajó Island	
Acarí	Vila Socorro, Lago Grande da Franca, Pará; Ipaupixuna near Santarém, Pará	Urucurituba near Alenquer, Pará
Achada	Arapiuns River, Pará; Vila Socorro, Lago Grande da Franca, Pará	
Amarelinha	Km 55 of Juriti-Tabatinga road, Pará	Piracauera da Cima near Santarém, Pará (2); Surubim-Açu near Santarém, Pará (2)
Anta	Jambu Açu, PA 252 highway between Moju and Acará, Pará	
Apoyona	Boa Esperança, km 45 of Santarém–Curuá-Una highway, Pará	
Arara amarela	Caxiuanã, Pará	
Bacuri	Caxiuanã, Pará	
Bemtevi	Boa Esperança, km 45 of Santarém–Curuá-Una highway, Pará	
Boi	Andirobalzinho, Santarém–Alter do Chão highway, Pará; Belterra, Pará; Ipaupixuna near Santarém, Pará	
Boi branco		Igarapé Jari near Arapixuna, Pará
Bragança	Jambu Açu, PA 252 highway, between Moju and Acará, Pará	

(*continued*)

Variety	Upland Location(s)	Amazon Floodplain Location(s)
Branquinha	Vila Forquilha near Tomé-Açu, Pará; km 56 of Santarém-Ruropólis road, Pará	Urucurituba near Alenquer, Pará
Brebe	Lago do Veado, mouth of Arapiuns near Santarém, Pará	
Bujaru	São João near Capitão Poço, Pará	
Burrona	Murumuru near Santarém, Pará (2)	
Cacao	Viçinal Ferrovia, km 35 of PA 150 highway, south of Marabá, Pará	
Camarão	Ramal São Pedro, km 115 of BR 316 highway, Belém-Brasília, Pará	
Capoeira	Lago do Veado, mouth of Arapiuns near Santarém, Pará	
Caratinga	Arapiuns River, Pará; Vila Socorro, Lago Grande da Franca, Pará	
Carema	Viçinal Ferrovia, km 35 of PA 150 highway, south of Marabá, Pará	
Cearense	Vila Forquilha near Tomé-Açu, Pará	
Chapeu de sol	Jambu Açu, PA 252 highway, between Moju and Acará, Pará	
Chavita	Murumuru near Santarém, Pará (3); Ipaupixuna near Santarém, Pará	
Consolada	Jambu Açu, PA 252 highway, between Moju and Acará, Pará	
Corací	Matá, Igarapé do Mamauru, near Óbidos, Pará; Flexal, Pará	
Coroazinho	Arapiuns River, Pará	
Curueira	Caxiuanã, Pará	
Dorotea		Ilha Tapará, near Santarém, Pará; Piracauera de Cima near Santarém, Pará
Douradinha	Aninduba, Pará	
Flor de boi	Arapixuna, Pará; Ipaupixuna near Santarém, Pará	Piracauera da Cima near Santarém, Pará; Urucurituba near Alenquer, Pará; Igarapé Jari near Arapixuna, Pará; Surubim-Açu near Santarém, Pará (2)
Gordura	São João near Capitão Poço, Pará	Januaria, Ilha do Carmo near Alenquer, Pará
Inajá	Flexal, Pará	Urucurituba near Alenquer, Pará
Inajázinha	Lastancia near Itupiranga, Pará	
Jaibara	Del Rey near Paragominas, Pará; Viçinal Ferrovia, km 35 of PA 150 highway, south of Marabá, Pará; Lastancia near Itupiranga, Pará	
Jibarinha	São João Batista near Itupiranga, Pará	
Juriti	Lastancia near Itupiranga, Pará	
Mãe Luisa		Igarapé Jari near Arapixuna, Pará
Mamão		Urucurituba near Alenquer, Pará
Mandé-miri	Caxiuanã, Pará	

Variety	Upland Location(s)	Amazon Floodplain Location(s)
Manivão	Ramal São Pedro, km 115 of BR 316 highway, Belém-Brasília, Pará	
Marapanim	Vila Forquilha near Tomé-Açu, Pará	
Maraquinha	Jambu Açu, PA 252 highway, between Moju and Acará, Pará	
Milagrosa	Boa Esperança, km 45 of Santarém–Curuá-Una highway, Pará; Murumuru near Santarém, Pará	
Mulatinha	Tomé-Açu, Pará	Urucurituba near Alenquer, Pará
Nambu	Vila Socorro, Lago Grande da Franca, Pará	
Nambuzinho	Km 55 of Juriti-Tabatinga road, Pará	
Pacajá	Vila Forquilha near Tomé-Açu, Pará; near Rio Camará, Marajó Island	Urucurituba near Alenquer, Pará
Pai Lourenço	Flexal, Pará	
Paruí	Near Rio Camará, Marajó Island	
Pau grande	Caxiuanã, Pará	
Pequi	São João near Capitão Poço, Pará	
Perereca	Arapiuns River, Pará; Vila Socorro, Lago Grande da Franca, Pará	
Piraíba		Urucurituba near Alenquer, Pará
Pororoca	Flexal, Pará; km 55 of Juriti-Tabatinga road, Pará	
Preta	Vila Socorro, Lago Grande da Franca, Pará; Comunidade São João near Itupiranga, Pará	
Pretinha	Murumuru near Santarém, Pará; Ipaupixuna near Santarém, Pará; km 16 of Morada Nova–Moju stretch of PA 150 highway, near Marabá, Pará; Rio Camará, Marajó Island	
Preta rica	Km 165 of Altamira-Uruará, Transamazon Highway, Pará	
Quinia	São João near Capitão Poço, Pará	
Roxinha	Arapixuna, Pará; Aninduba, Pará; Lago do Veado, mouth of Arapiuns near Santarém, Pará; Vila Socorro, Lago Grande da Franca, Pará	Surubim-Açu near Santarém, Pará
São José	Vila Socorro, Lago Grande da Franca, Pará	
Sapinha	Arapixuna, Pará	
Seis meses	Andirobalzinho, Santarém–Alter do Chão highway, Pará	
Sete Meses	Jacaré-Capá near Monte Alegre, Pará	
Sutinga	Del Rey near Paragominas, Pará	
Tachi amarela	Del Rey near Paragominas, Pará	
Tachi branca	Del Rey near Paragominas, Pará (3)	
Tambaqui		Boca Aritapera de Cima near Santarém, Pará

(*continued*)

Variety	Upland Location(s)	Amazon Floodplain Location(s)
Tapajós	Lago do Veado, mouth of Arapiuns near Santarém, Pará	
Tartaruga	Arapixuna, Pará	Near Arapixuna, Pará
Tartaruginha	Aninduba, Pará	
Torrão	São Pedro, km 115 of BR 316 highway, Belém-Brasília, Pará	
Tres meses	Gleba Pombal near Juscimeira, Mato Grosso	
Tucumã	Murumuru near Santarém, Pará	
Urucuri	Viçinal Ferrovia, km 35 of PA 150 highway south of Marabá, Pará; km 46 of Santarém-Rurópolis highway, Pará	
Vassourinha	Gleba Pombal near Juscimeira, Mato Grosso	
Vermelhinha	Caxiuanã, Pará	
Zucuda	Vila Socorro, Lago Grande da Franca, Pará	

Note: Numbers in parentheses indicate occurrences in different fields at that location.

Source: Field notes 1992–1997.

Appendix D

Twenty-two Varieties of Sweet Manioc Cultivated in the Brazilian Amazon

Variety	Upland Location(s)	Amazon Floodplain Location(s)
Amarelinha	Agrovila Nova Fronteira, km 80 of Transamazon Highway, Pará; Flexal, Pará	Piracauera de Cima near Santarém, Pará
Americana	Del Rey near Paragominas, Pará	
Boliviana	Km 56 of Santarém-Rurópolis highway, Pará	
Branca	Murumuru, near Santarém, Pará; Arapixuna, Pará (2); Lago do Veado, mouth of Arapiuns near Santarém, Pará	Urucurituba near Alenquer, Pará; Igarapé Jari, near Arapixuna, Pará; Afuá, Marajó Island
Cacao	Del Rey near Paragominas, Pará; Machadinho, Rondônia; Quatro Marcos, Mato Grosso	
Casca roxa	Quatro Marcos, Mato Grosso	
Casca vermelha	Murumuru near Santarém, Pará	Urucurituba near Alenquer, Pará
Cayena	Del Rey, Municipality of Paragominas, Pará	
Da gorda		Urucurituba near Alenquer, Pará
Liberata	Gleba Pombal near Juscimeira, Mato Grosso; Araputanga, Mato Grosso	
Manteiga	Rio Camará, Marajó Island; Ipaupixuna near Santarém, Pará	Urucurituba near Alenquer, Pará
Manteiginha	Del Rey near Paragominas, Pará	Afuá, Marajó Island
Meninha		Urucurituba near Alenquer, Pará
Preta	Belterra, Pará; Lastancia near Itupiranga, Pará; São João Batista near Itupiranga, Pará	Urucurituba near Alenquer, Pará; Igarapé Jari, near Arapixuna, Pará; Ituqui near Santarém, Pará

(*continued*)

Variety	Upland Location(s)	Amazon Floodplain Location(s)
Santa Rosa	Igarapé-Açu, Pará	
Rosa	Jacaré-Capá near Monte Alegre, Pará	Ituqui near Santarém, Pará
Roxa	Lago do Veado, mouth of Arapiuns near Santarém, Pará	
Uruim	Del Rey near Paragominas, Pará; Viçinal Ferrovia, km 35 of PA 150 highway, south of Marabá, Pará	
Tiririca	Del Rey near Paragominas, Pará; km 46 of Santarém-Rurópolis highway, Pará	
Vassourinha	Km 165 of Altamira-Uruará, Transamazon Highway, Pará; Quatro Marcos, Mato Grosso; Araputanga, Mato Grosso	
Vermelha	Belterra, Pará; Ipaupixuna near Santarém, Pará	
Vermelhinha	Km 55 of Juriti-Tabatinga road, Pará	

Note: Numbers in parentheses indicate occurrences in different fields at that location.

Source: Field notes 1992–1997.

Appendix E

Ninety-three Plant Species Growing in Twenty-two Home Gardens on the Amazon Floodplain

Name	Home Gardens	Occurs Wild on Floodplain
Açaí palm	2	+
Ambarella	1	
Andirá-uxi	2	+
Annatto	2	
Apé	1	+
Apuí	1	+
Arabica coffee	1	
Araçá	3	+
Aruanãzeiro	3	+
Avocado	3	
Azeitona (unidentified)	1	
Bacaba palm	2	
Bacuri	3	+
Baianeiro	1	+
Banana	18	
Biribá	1	
Boteiro	2	+
Breadfruit	1	
Buriti palm	2	+
Bussú palm	1	+
Cacao	4	
Caçeira	1	
Calabash gourd	18	
Cannonball tree	2	+
Carambola	1	
Carnaúba palm	2	
Cashew	9	
Catauarí	6	+
Caxinguba	1	+
Cecropia	1	+
Coconut palm	10	
Common bean	1	
Cotton	3	
Cupuaçu	6	
Curumi	2	+
Cutitiribá	1	
Genipap	7	
Guava	14	
Ingá	3	
Ingá baú	1	+
Ingá capuchino	2	
Ingá cipó	3	+
Ingá grande	1	+
Ingá xixi	1	
Jambolan	1	

(*continued*)

Name	Home Gardens	Occurs Wild on Floodplain	Name	Home Gardens	Occurs Wild on Floodplain
Jucá	3		Pitomba	1	
Lime	7		Pojó	2	
Maize	1		Rubber	2	
Malay apple	3		Runner bean	2	
Mango	15		Sapodilla	1	
Manioc	3		Sapucaia	5	+
Marimari	3	+	Sapupira	2	+
Marinemera (unidentified)	1		Socoró	3	+
			Soursop	6	
Marisara (unidentified)	4	+	Squash	1	
Meracruera (unidentified)	2	+	Sugarcane	5	
			Surinam cherry	1	
Mulato wood	4	+	Sweet orange	6	
Munguba	5	+	Sweet potato	3	
Muruci	1		Sweetsop	1	
Murumuru palm	1	+	Tachi	1	+
Papaya	3		Tamarind	2	
Paricá da várzea	1	+	Tangerine	1	
Passionfruit	3		Tarumã	3	+
Patazana	2	+	Tucumã palm	1	
Pau de Angola	2		Uruá	9	+
Peach palm	1		Urucuri palm	1	+
Pião branco	4		Vassoura	1	
Pião roxo	2		Watermelon	2	
Pineapple	2		Yellow mombim	5	+

Notes: This list is exclusive of flowers, herbaceous medicinal plants, spices, and vegetables.

Home gardens were inventoried at the following locations: Four on Careiro Island near Manaus, Amazonas, in 1972; seventeen along the Juriti-Santarém stretch of the river in Pará between 1992 and 1996; and one near Afuá on Marajó Island in 1997.

Notes

Chapter 2

1. Cook, O. F., *Vegetation affected by agriculture in Central America*, Bulletin of Plant Industry, no. 145 (U.S. Department of Agriculture, Washington, D.C., 1909), p. 12.

2. See, for example, Dillehay, T. D., and M. B. Collins, "Early cultural evidence from Monte Verde in Chile," *Nature* 332 (1988): 150–52. Debates about the antiquity of man in the Americas have often been sharp, with many reluctant to accept the idea that people have been moving into South America in waves since more than ten thousand years ago. Land and ice bridges joined Alaska with Asia several times in the Pleistocene, and small groups of hunters and gatherers took advantage of them to settle the New World.

3. Early dates for arrival of humans in South America have often been contested on the grounds that alleged occupation sites were disturbed and thus contaminated with older material. And there has been a reluctance to accept the fact that new stone-working styles arose in South America, often different from Clovis points found in the United States. Modern DNA tests (see Gibbons, A., "Geneticists trace the DNA trail of the first Americans," *Science* 259 [1993]: 312–13) and sophisticated linguistic analysis (see Horgan, J., "Early arrivals: scientists argue over how old the New World is," *Scientific American* 266.2 (1992): 17–20) both point to the arrival of people in South America tens of thousands of years ago.

4. Only one indigenous group in Amazonia reportedly lives in the forest canopy today, the uncontacted Piriutiti, who live in the Waimiri-Atroari reserve near the border between Amazonas and Roraima; that group lives in the treetops as a defensive strategy. See Ricardo, C. A., *Povos indígenas no Brasil 1991/1995* (Instituto Sociambental, São Paulo, 1996). But canopy dwelling may have been more common in the distant past.

5. Roosevelt, A. C., M. L. Costa, C. L. Lopes, M. Michab, N. Mercier, H. Valladas, J. Feathers, W. Barnett, M. I. Silveira, A. Henderson, J. Silva, B. Chernoff, D. S. Reese, J. A. Holman, N. Toth, and K. Schick, "Paleoindian cave dwellers in the Amazon: The peopling of the Americas," *Science* 272 (1996): 373–84.

6. Roosevelt, A. C., R. A. Housley, M. Imazio da Silveira, S. Maranca, and R. Johnson, "Eighth millenium pottery from a prehistoric shell midden in the Brazilian Amazon," *Science* 254 (1991): 1621–24.

7. Meggers, B. J., *Man and culture in a counterfeit paradise* (Aldine, Chicago, 1971).

8. Mora, S., "Cold and hot, green and yellow, dry and wet: Direct access to resources in Amazonia and the Andes," *Florida Journal of Anthropology* 18 (1993): 51–60.

9. Meggers, B. J., "Amazonia on the eve of European contact: Ethnohistorical, ecological, and anthropological perspectives," *Revista de Arqueología Americana* 8 (1995): 91–115. Betty Meggers has remained an inveterate environmental determinist throughout her productive career, and we are indebted to her for stimulating so much debate and research on human carrying capacity in Amazonia.

10. Cook, S. F., and W. Borah, *Essays in population history*, 3 vols. (University of California Press, Berkeley, 1971 [vol. 1], 1974 [vol. 2], 1979 [vol. 3]). The historians Cook and Borah used meticulous analysis of colonial and church records to arrive at higher numbers than had hitherto been suspected for indigenous people in Central America and the Caribbean at time of contact with Europeans.

11. The Portuguese seafarer Pedro Álvares Cabral landed on the coast of northeastern Brazil in 1500, apparently blown off course near South Africa on his way from Lisbon to the Orient, but he only spent a few days in Brazil and never reached the Amazon. Other Portuguese voyagers may have brushed the mouth of the Amazon in the early 1500s, but they appear to have had little if any contact with indigenous groups.

12. Medina, J. T., *The discovery of the Amazon* (Dover, New York, 1998).

13. DeBoer, W. R., "Buffer zones in the cultural ecology of aboriginal Amazonia: An ethnohistorical approach," *American Antiquity* 48.2 (1981): 364–77.

14. Brown, C. B., and W. Lidstone, *Fifteen thousand miles on the Amazon and its tributaries* (Edward Stanford, London, 1878), p. 271.

15. For the black-earth site at Santarém, see Roosevelt, A. C., "Resource management in Amazonia before the conquest: Beyond ethnographic projection," *Advances in Economic Botany* 7 (1989): 30–62. For other anthrosol sites in the vicinity of the confluence of the Amazon and the Tapajós, see Nimuendajú, C., "Os Tapajó," *Boletim do Museu Goeldi* 10 (1949): 93–106.

16. Denevan, W. M., "A bluff model of riverine settlement in Prehistoric Amazonia," *Annals of the Association of American Geographers* 86.4 (1996): 654–81.

17. Lathrap, D. W., "Aboriginal occupation and changes in river channel on the central Ucayali, Peru," *American Antiquity* 33.1 (1968): 62–79.

18. Sternberg discusses anthrosols on the Amazon floodplain in *A Água e Homem na Várzea do Careiro* (Universidade do Brasil, Rio de Janeiro, 1956) and in "Radiocarbon dating as applied to a problem of Amazonian morphology," in *Comptes Rendus du XVIII Congrès International de Géographie, Rio de Janeiro, 1956* (Centro de Pesquisas de Geografia do Brasil, Universidade do Brasil, Rio de Janeiro, 1960), pp. 399–424.

19. Roosevelt, A. C., *Moundbuilders of the Amazon: Geophysical archaeology on Marajó Island, Brazil* (Academic Press, San Diego, 1991).

20. Sauer, C. O., *Agricultural origins and dispersals: The domestication of animals and foodstuffs* (MIT Press, Cambridge, Massachusetts, 1969). Sauer, a Berkeley geographer, was a pioneer thinker in turning our attention to the tropics as the venue for the earliest agriculture, especially in terms of root crops.

21. See Roosevelt, *Moundbuilders of the Amazon.*

22. Sampaio, F. X. R., *Diário da viagem que em visita, e correição, das povações da capitania de S. Joze do Rio Negro fez o ouvidor, e intendente geral da mesma, no anno de 1774 e 1775* (Typographia da Academia, Lisbon, 1825), p. 73. Sampaio was a traveling judge and an astute observer of economic activities in the Brazilian Amazon during the late 1700s.

23. Ibid.

24. Spruce, R., *Notes of a botanist on the Amazon and Andes*, 2 vols. (Macmillan, London, 1908), vol. 2, p. 176.

25. Bates, H. W., *The naturalist on the river Amazons*, 2 vols. (John Murray, London, 1863), vol. 1, p. 22.

26. Smith, H. H., *Brazil: The Amazons and the coast* (Charles Scribner's Sons, New York, 1879), p. 266.

27. Le Cointe, P., "Le climat Amazonien, et plus spécialement le climat du Bas Amazone," *Annales de Géographie* 15 (1906): 449–62.

28. Edmundson, G., *Journal of the travels and labours of Father Samuel Fritz in the river of the Amazons between 1686 and 1723* (Hakluyt Society, London, 1922), p. 71.

29. Jobim, A., *Monografia geográfica do Estado do Amazonas* (Papelaria Velho Lino, Manaus, 1949).

30. See Sampaio, *Diário da viagem que em visita*, p. 22.

31. See Brown and Lidstone, *Fifteen thousand miles on the Amazon*, p. 98.

32. See Bates, *The naturalist on the river Amazons*, vol. 2, p. 157.

33. Agassiz, L., *A journey in Brazil* (Houghton, Mifflin, Boston, 1896), p. 228.

Chapter 3

1. Le Cointe, P., *Árvores e plantas úteis* (Editora Nacional, Rio de Janeiro, 1947).

2. Anderson, A. B., and E. M. Ioris, "Valuing the rain forest: Economic strategies by small-scale forest extractivists in the Amazon estuary," *Human Ecology* 20.3 (1992): 337–69; Anderson, A. B., A. Gely, J. Strudwick, G. L. Sobel, and M. G. C. Pinto, "Um sistema agroflorestal na várzea do estuário Amazônico (Ilha das Onças, Município de Bacarena, Estado do Pará)," *Acta Amazonica*, Supl. 15.1–2 (1985): 195–224.

3. Agassiz, *A journey in Brazil*, pp. 228, 140.

4. Frikel, P., "Áreas de arboricultura pré-agrícola na Amazônia: Notas preliminares," *Revista de Antropologia* 21.1 (1978): 45–52.

5. Chandless, W., "Ascent of the river Purus," *Journal of the Royal Geographical Society* 36 (1866): 86–118.

6. Huber, J., "Mattas e madeiras amazonicas," *Boletim do Museu Goeldi de Historia Natural e Ethnographia* 6 (1910): 91–225.

7. Schultes, R. E., "De plantis toxicariis e Mundo Novo tropicale commentationes XXXII: Notes, primarily of field tests and native nomenclature, on biodynamic plants of the northwest Amazon," *Botanical Museum Leaflets* (Harvard) 29.3 (1983): 251–72.

8. Spix, J. B., and C. F. P. Martius, *Viagem pelo Brasil, 1817–1820*," vol. 3 (Edições Melhoramentos, São Paulo, 1976), p. 83.

9. Gates, R. R., *A botanist in the Amazon valley: An account of the flora and fauna in the land of floods* (H. F. & G. Witherby, London, 1927), p. 49.

10. Goulding, M., *The fishes and the forest: Explorations in Amazonian natural history* (University of California Press, Berkeley, 1980).

11. Camargo, F. C., "Report on the Amazon region," in *Problems of humid tropical regions* (UNESCO, Paris, 1958), pp. 11–24; Woodroffe, J. F., and H. H. Smith, *The rubber industry of the Amazon and how its supremacy can be maintained* (John Bale, Sons & Danielsson, London, 1915).

12. Pendelton, L. H., "Trouble in paradise: Practical obstacles to nontimber forestry in Latin America," in *Sustainable harvest and marketing of rain forest products*, edited by M. Plotkin and L. Famolare (Island Press, Washington, D.C., 1992), pp. 252–62.

Chapter 4

1. Mathews, E. D., *Up the Amazon and Madeira rivers, through Bolivia and Peru* (Sampson, Low, Marston, Searle and Rivington, London, 1879), p. 9.

2. Maw, H. L., *Journal of a passage from the Pacific to the Atlantic, crossing the Andes and descending the River Marañon or Amazon* (John Murray, London, 1829), p. 284.

3. Smith, *Brazil*, p. 95.

4. Furtado, L. G., *Pescadores do Rio Amazonas: Um estudo antropológico da pesca ribeirinha numa área Amazônica* (Museu Paraense Emílio Goeldi, Belém, 1993), p. 174.

5. Sternberg, *A Água e Homem na várzea do Careiro*, p. 165.

6. Bates, *The naturalist on the river Amazons*, vol. 2, p. 37.

7. Smyth, W., and F. Lowe, *Narrative of a journey from Lima to Pará, across the Andes and down the Amazon* (John Murray, London, 1836), p. 301.

8. Anderson, A. B., and E. M. Ioris, "Valuing the rain forest: Economic strategies by small-scale forest extractivists in the Amazon estuary," *Human Ecology* 20.3 (1992): 337–69.

Chapter 5

1. *O Liberal* (Belém, Pará, Brazil), 11 November 1993, *Dia a Dia* section, p. 3.
2. The employer customarily provides lunch and coffee to the workers. The latter can negotiate a slightly higher daily wage if they bring their own food and coffee.
3. Brigham, W. T., *Guatemala, the land of the quetzal* (Charles Scribner's Sons, New York, 1887), p. 39.

Chapter 6

1. Harris, D. R., "The agricultural foundations of lowland Maya civilization: A critique," in *Pre-Hispanic Maya agriculture*, edited by P. D. Harrison and B. L. Turner (University of New Mexico Press, Albuquerque), pp. 301–22; Rindos, D., *The origins of agriculture: An evolutionary perspective* (Academic Press, Orlando, 1984).
2. Bates, *The naturalist on the river Amazons*, vol. 1, p. 323.

Chapter 7

1. Hiraoka, M., *Japanese agricultural settlement in the Bolivian Upper Amazon: A study in regional ecology*, Latin American Studies, no. 1, Special Research Project on Latin America (University of Tsukuba, Sakuramura, Ibaraki, Japan, 1980).
2. Suspicions about the underlying intentions of the Hudson Institute's proposal to dam the Amazon can be found in Valverde, O., "Dos grandes lagos sul-Americanos aos grandes eixos rodoviários," *Caderno de Ciências da Terra* (Instituto de Geografia, Universidade de Sao Paulo) 14 (1971): 1–22; Reis, A. C. F., "Brasileiros e estrangeiros na ocupação da Amazônia," *A Amazônia Brasileira em Foco* 5 (1971): 7–15.

Further Reading

Chapter 1

On links between biodiversity and agriculture

Pimentel, D., U. Stachow, D. A. Takacs, H. W. Brubaker, A. R. Dumas, J. J. Meaney, J. A. S. O'Neil, D. E. Onsi, and D. B. Corzilius. 1992. Conserving biological diversity in agricultural/forestry systems. *Bioscience* 42:354–62.

Smith, N. J. H. 1996. Effects of land-use systems on the use and conservation of biodiversity. In *Biodiversity and agricultural intensification: Partners for development and intensification*, edited by J. P. Srivastava, N. J. H. Smith, and D. Forno, pp. 52–79. Environmentally Sustainable Development Studies and Monographs Series, no. 11, World Bank, Washington, D.C.

Srivastava, J., N. J. H. Smith, and D. Forno. 1996. *Biodiversity and agriculture: Implications for conservation and development*. World Bank, Technical Paper 321, Washington, D.C.

On the cultural dimensions to supposedly natural habitats

Balée, W. 1989. The culture of Amazonian forests. *Advances in Economic Botany* 7:1–21.

Balée, W., and A. Gély. 1989. Managed forest succession in Amazonia: The Ka'apor case. *Advances in Economic Botany* 7:129–58.

Barcelos, E. 1986. *Características genético-ecológicas de populações naturais de caiaué (Elaeis oleifera, H. B. K., Cortés) na Amazônia brasileira*. INPA, Manaus.

Cook, O. F. 1909. *Vegetation affected by agriculture in Central America*. Bulletin of Plant Industry, no. 145, U.S. Department of Agriculture, Washington, D.C.

———. 1921. Milpa agriculture, a primitive tropical system. In *Annual report of the board of regents*, pp. 307–26. Smithsonian Institution, Washington, D.C.

Coomes, O. T. 1995. A century of rain forest use in western Amazonia: Lessons for extraction-based conservation of tropical forest resources. *Forest and Conservation History* 39:108–20.

De Blank, S. 1952. A reconnaissance of the American oil palm *Elaeis melanococca* (Gaertner em. Bailey) = *Corozo oleifera* (Giseke), *Alfonsia oleifera* (H. B. K.). *Tropical Agriculture* 29:90–101.
Dufour, D. L. 1990. Use of tropical rainforests by native Amazonians. *Bioscience* 40:652–59.
Hecht, S. B. 1992. Extractive communities, biodiversity and gender issues in Amazonia. In *Proceedings of the international conference on women and biodiversity* (Kennedy School of Government, Harvard University, October 4–6, 1991), edited by L. M. Borkenhagen and J. N. Abromovitz, pp. 52–63. Cambridge: Committee on Women and Biodiversity, Harvard University, Cambridge, Massachusetts.
Leavitt, F. J. 1992. The case for conservation. *Nature* 360:100.
Parsons, J. J. 1986. "Now this matter of cultural geography": Notes from Carl Sauer's last seminar at Berkeley. In *Carl O. Sauer: A tribute*, edited by M. Kenzer, pp. 153–63. Oregon State University Press, Corvalis.
Sauer, J. D. 1988. *Plant migration: The dynamics of geographic patterning in seed plant species*. University of California Press, Berkeley.
Siskind, J. 1973. *To hunt in the morning*. Oxford University Press, London.
Smith, N. J. H. 1995. Human-induced landscape changes in Amazonia and implications for development. In *Global land use change: A perspective from the Columbian encounter*, edited by B. L. Turner, II, A. Gómez, Fernando González, and F. Castri, pp. 221–51. Consejo Superior de Investigaciones Científicas, Madrid.
Toledo, V. M., B. Ortiz, and S. Medellín. 1994. Biodiversity islands in a sea of pasturelands: Indigenous resource management in the humid tropics of Mexico. *Etnoecológica* 2.3:37–49.
Turner, II, B. L., and K. W. Butzer. 1992. The Columbian encounter and land-use change. *Environment* 34.8:16–20, 37–44.

On natural history and cultural ecology of the Amazon River

Denevan, W. M. 1984. Ecological heterogeneity and horizontal zonation of agriculture in the Amazon floodplain. In *Frontier expansion in Amazonia*, edited by M. Schmink and C. H. Wood, pp. 311–36. University of Florida Press, Gainesville.
Goulding, M. 1980. *The fishes and the forest: Explorations in Amazonian natural history*. University of California Press, Berkeley.
———. 1989. *Amazon: The flooded forest*. BBC Books, London.
———. 1993. Flooded forests of the Amazon. *Scientific American* 266.3:114–20.
Goulding, M., N. J. H. Smith, and D. Mahar. 1996. *Floods of fortune: Ecology and economy along the Amazon*. Columbia University Press, New York.
Hiraoka, M. 1985. Changing floodplain livelihood patterns in the Peruvian Amazon. *Tsukuba Studies in Human Geography* 9:243–75.
Junk, W. J., ed. 1997. *The central Amazon floodplain: Ecology of a pulsing system*. Springer, Berlin.
Moran, E. F. 1993. *Through Amazonian eyes*. University of Iowa Press, Iowa City.
Sternberg, H. O'R. 1975. *The Amazon River of Brazil*. Franz Steiner, Wiesbaden.
Yuyama, K. 1993. Potencialidade agrícola dos solos da várzea e a utilização de leguminosas na Amazônia central. In *Bases científicas para estratégias de preservação e desenvolvimento da Amazônia*, edited by E. J. Ferreira, G. M. Santos, E. L. Leão, and L. A. Oliveira. Vol. 2, pp. 223–39. INPA, Manaus.

Chapter 2

On the notion that the Amazon has a limited capacity to support human populations

Burns, E. B. 1993. *A history of Brazil*. Columbia University Press, New York.
Hames, R. 1991. Wildlife conservation in tribal societies. In *Biodiversity: Culture, conservation, and ecodevelopment*, edited by M. L. Oldfield and J. B. Alcorn, pp. 172–99. Westview Press, Boulder, Colorado.

Meggers, B. J. 1971. *Man and culture in a counterfeit paradise.* Aldine, Chicago.
———. 1995. Amazonia on the eve of European contact: Ethnohistorical, ecological, and anthropological perspectives. *Revista de Arqueología Americana* 8:91–115.

On the significance of geometric designs in ancient pictographs

Bower, B. 1996. Visions on the rocks: Rock and cave art may offer insights into shamans' trance states and spiritual sightings. *Science News* 150:216–17.

On the paleoecology and prehistory of tropical America

Appenzeller, T. 1992. A high five from the first New World settlers? Bones, stones, and a palm print from a New Mexico cave could shake up New World archaeologists. *Science* 255:920–21.

Butzer, K. W. 1991. An Old World perspective on potential mid-Wisconsinan settlement in the Americas. In *The first Americans: Search and research*, edited by T. D. Dillehay and D. J. Metzer, pp. 137–56. CRC Press, Boca Raton, Florida.

Colinvaux, P. A., and M. B. Brush. 1991. The rain-forest ecosystem as a resource for hunting and gathering. *American Anthropologist* 93.1:153–60.

Cook, S. F., and W. Borah. 1971. *Essays in population history*. Vol. 1, *Mexico and the Caribbean*. University of California Press, Berkeley.

———. 1974. *Essays in population history*. Vol. 2, *Mexico and the Caribbean*. University of California Press, Berkeley.

———. 1979. *Essays in population history*. Vol. 3, *Mexico and the California*. University of California Press, Berkeley.

Dillehay, T. D., and M. B. Collins. 1988. Early cultural evidence from Monte Verde in Chile. *Nature* 332:150–52.

Gibbons, A. 1993. Geneticists trace the DNA trail of the first Americans. *Science* 259:312–13.

Guidon, N., and G. Delibrias. 1986. Carbon-14 dates point to man in the Americas 32,000 years ago. *Nature* 321:769–71.

Horgan, J. 1992. Early arrivals: Scientists argue over how old the New World is. *Scientific American* 266.2:17–20.

Meltzer, D. J., J. M. Adovasio, and T. D. Dillehay. 1994. On a Pleistocene human occupation at Pedra Furada, Brazil. *Antiquity* 68:695–714.

Sauer, C. O. 1969. *Agricultural origins and dispersals: The domestication of animals and foodstuffs*. MIT Press, Cambridge, Massachusetts.

On the paleoecology and prehistory of Amazonia

Carneiro, R. L. 1970. Transition from hunting to horticulture in the Amazon Basin. In *Proceedings of the 8th congress of anthropological and ethnological sciences, Tokyo and Kyoto, 1868*. Vol. 3, *Ethnology and Archaeology*, pp. 244–48.

Comas, J. 1951. La realidad del trato dado a los indígenas de América entre los siglos XV y XX. *América indígena* 11.4:323–70.

DeBoer, W. R. 1981. Buffer zones in the cultural ecology of aboriginal Amazonia: An ethnohistorical approach. *American Antiquity* 48.2:364–77.

Denevan, W. M. 1996. A bluff model of riverine settlement in prehistoric Amazonia. *Annals of the Association of American Geographers* 86.4:654–81.

Erickson, C. L. 1995. Archaeological methods for the study of ancient landscapes of the Llanos de Mojos in the Bolivian Amazon. In *Archaeology in the lowland American tropics: Current analytical methods and applications*, edited by P. W. Stahl, pp. 66–95. Cambridge University Press, Cambridge.

Ferreira Penna, D. S. 1876. Breve notícia sôbre os sambaquis do Pará. *Arquivos do Museu Nacional* 1:85–99.

Hartt, C. F. 1885. Contribuições para a ethnologia do valle do Amazonas. *Archivos do Museu Nacional* 6:1–174.

Hemming, J. 1978. *Red gold: The conquest of the Brazilian indians, 1500–1760*. Harvard University Press, Cambridge, Massachusetts.

Heriarte, M. 1964. *Descriçam do Estado do Maranham, Para, Corupa, Rio das Amazonas (1662)*. Akademische Druck und Verlagsanstalt, Graz, Austria.

Irion, G. 1989. Quaternary geological history of the Amazon lowlands. In *Tropical forests: Botanical dynamics, speciation and diversity*, edited by L. B. Holm-Nielsen, I. C. Nielsen, I. C., and H. Balslev, pp. 23–34. Academic Press, London.

Kern, D. C., and N. Kämpf. 1989. Antigos assentamentos indígenas na formação de solos com terra preta arqueológica na região de Oriximiná, Pará. *Revista Brasileira de Ciências de Solo* 13:219–25.

Lathrap, D. W. 1968. Aboriginal occupation and changes in river channel on the central Ucayali, Peru. *American Antiquity* 33.1:62–79.

———. 1977. Our father the cayman, our mother the gourd: Spinden revisited, or a unitary model for the emergence of agriculture in the New World. In *Origins of agriculture*, edited by C. A. Reed, pp. 713–51. Mouton, The Hague.

Le Cointe, P. 1945. *O Estado do Pará: A terra, a água e o ar*. Editora Nacional, São Paulo.

Medina, J. T. 1988. *The discovery of the Amazon*. Dover, New York.

Mora, S. 1993. Cold and hot, green and yellow, dry and wet: Direct access to resources in Amazonia and the Andes. *Florida Journal of Anthropology* 18:51–60.

Nimuendajú, C. 1949. Os Tapajó. *Boletim do Museu Goeldi* 10:93–106.

Roosevelt, A. C. 1989. Resource management in Amazonia before the conquest: Beyond ethnographic projection. *Advances in Economic Botany* 7:30–62.

———. 1991. *Moundbuilders of the Amazon: Geophysical archaeology on Marajó Island,* Brazil. Academic Press, San Diego.

———. 1992. Secrets of the forest: An archaeologist reappraises the past—and future—of Amazonia. *The Sciences* 32.6:22–28.

Roosevelt, A. C., R. A. Housley, M. Imazio da Silveira, S. Maranca, and R. Johnson. 1991. Eighth millenium pottery from a prehistoric shell midden in the Brazilian Amazon. *Science* 254:1621–24.

Roosevelt, A. C., M. L. Costa, C. L. Lopes, M. Michab, N. Mercier, H. Valladas, J. Feathers, W. Barnett, M. I. Silveira, A. Henderson, J. Silva, B. Chernoff, D. S. Reese, J. A. Holman, N. Toth, and K. Schick. 1996. Paleoindian cave dwellers in the Amazon: The peopling of the Americas. *Science* 272:373–84.

Simões, M. F. 1976. Nota sobre duas pontas-de-projétil da Bacia do Tapajós (Pará). *Boletim do Museu Paraense Emílio Goeldi*, Nova Série 62:1–14.

Sioli, H. 1956. O Rio Arapiuns. *Boletim Técnico do Instituto Agronômico do Norte* 32:5–116.

Smith, N. J. H. 1980. Anthrosols and human carrying capacity in Amazonia. *Annals of the Association of American Geographers* 70:553–66.

Sternberg, H. O'R. 1956. *A Água e Homem na Várzea do Careiro*. Universidade do Brasil, Rio de Janeiro.

———. 1960. Radiocarbon dating as applied to a problem of Amazonian morphology. In *Comptes Rendus du 18 Congrès International de Géographie, Rio de Janeiro, 1956*, pp. 399–424, Centro de Pesquisas de Geografia do Brasil Universidade do Brasil, Rio de Janeiro.

Sweet, D. G. 1974. *A rich realm of nature destroyed: The middle Amazon valley, 1640–1750*. Ph.D. diss., University of Wisconsin.

On the settlement patterns and subsistence of indigenous groups and contemporary peasants

Acuña, C. 1891. *Nuevo descubrimento del gran rio de las Amazonas en el año 1639*. García, Madrid.

Chandless, W. 1866. Ascent of the River Purus. *Journal of the Royal Geographical Society* 36:86–118.

Marajó, J. C. 1992. *As regiões amazônicas: Estudos chorographicos dos estados do Gram Pará e Amazônas*. Secretaria de Estado da Cultura, Belém.

Métraux, A. 1963. Tribes of the Juruá-Purús basins. In *Handbook of South American indians*, edited by J. H. Steward, pp. 657–86. Cooper Square Publishers, New York.
Moran, E. F. 1990. *A ecologia humana das populações da Amazônia*. Vozes, Petrópolis.
Ricardo, C. A. 1996. *Povos indígenas no brasil 1991/1995*. Instituto Sociambental, São Paulo.
Smith, N. J. H. 1974. Destructive exploitation of the South American river turtle. *Yearbook of the Pacific Coast Geographers* 36:85–102.
———. 1981. *Man, fishes, and the Amazon*. Columbia University Press, New York.

On the diffusion of maize in South America and its implications for settlement

Bush, M. A., D. R. Piperno, and P. A. Colinvaux. 1989. A 6,000-year history of Amazonian maize cultivation. *Nature* 340:303–5.
Herrera, L. F., I. Cavelier, C. Rodriguez, and S. Mora. 1992. The technical transformation of an agricultural system in the Colombian Amazon. *World Archaeology* 24.1:98–113.
Pearsall, D. M. 1992. The origins of plant cultivation in South America. In *The origins of agriculture: An international perspective*, edited by C. W. Cowan and P. J. Watson, pp. 173–205. Smithsonian Institution Press, Washington, D.C.
Roosevelt, A. C. 1980. *Parmana: Prehistoric maize and manioc subsistence along the Amazon and Orinoco*. Academic Press, New York.

On the implications of dense aboriginal settlements for conservation and transformation of landscapes

Balée, W. 1989. Cultura na vegetação da Amazônia brasileira. In *Biologia e Ecologia Humana na Amazônia: Avaliação e Perspectivas*, edited by W. A. Neves, pp. 95–109. Museu Paraense Emílio Goeldi, Belém.
Huber, J. 1906. Sur l'indigénat du *Theobroma cacao* dans les alluvions du Purus et sur quelques autres espècies du genre *Theobroma*. *Bulletin de l'Herbier Boissier* 6:272–74.
———. 1910. Mattas e madeiras amazonicas. *Boletim do Museu Goeldi de Historia Natural e Ethnographia* 6:91–225.
Sampaio, F. X. R. 1825. *Diário da viagem que em visita, e correição, das povações da Capitania de S. Joze do Rio Negro fez o ouvidor, e intendente geral da mesma, no anno de 1774 e 1775*. Typographia da Academia, Lisbon.
Smith, N. J. H. 1996. *The enchanted Amazon rain forest: Stories from a vanishing world*. University Press of Florida, Gainesville.
Smith, N. J. H., J. T. Williams, D. L. Plucknett, and J. P. Talbot. 1992. *Tropical forests and their crops*. Cornell University Press, Ithaca, New York.
Spruce, R. 1908. *Notes of a botanist on the Amazon and Andes*. Vol. 1. Macmillan, London.

On the role of human-set fires in altering the vegetation of Amazonia

Aubréville, A. 1961. *Étude écologique des principales formations végétales du Brésil et contribution à la connaissance des forêts de l'Amazonie brésilienne*. Centre Technique Forestier, Nogent-sur-Marne.
Eden, M. J., and D. F. M. McGregor. 1992. Dynamics of the forest-savanna boundary in the Rio Branco-Rupununi region of northern Amazonia. In *Nature and dynamics of forest-savanna boundaries*, edited by P. A. Furley, J. Proctor, and J. A. Ratter, pp. 77–89. Chapman and Hall, London.
Hopkins, B. 1992. Ecological processes at the forest-savanna boundary. In *Nature and dynamics of forest-savanna boundaries*, edited by P. A. Furley, J. Proctor, and J. A. Ratter, pp. 21–33. Chapman and Hall, London.
Le Cointe, P. 1906. Le climat Amazonien, et plus spécialement le climat du Bas Amazone. *Annales de Géographie* 15:449–62.
Macedo, D. S., and A. B. Anderson. 1993. Early ecological changes associated with logging in an Amazon floodplain. *Biotropica* 25.2:151–63.

Ratter, J. A. 1992. Transitions between cerrado and forest vegetation in Brazil. In *Nature and dynamics of forest-savanna boundaries*, edited by P. A. Furley, J. Proctor, and J. A. Ratter, pp. 417–29. Chapman and Hall, London.

Sanaiotti, T. N. 1991. Ecologia de paisagens: Savannas Amazônicas. In *Bases científicas para estratégias de preservação e desenvolvimento da Amazônia: Fatos e perspectivas*, edited by A. L. Val, R. Figliuolo, and E. Feldberg. Vol. 1, pp. 77–81. INPA, Manaus.

Scott, G. A. J. 1977. The role of fire in the creation and maintenance of savanna in the montaña of Peru. *Journal of Biogeography* 4:143–67.

On the postcontact population crash and the slave trade

Cruz, F. L. 1942. *Nuevo descubrimento del Rio de las Amazonas hecho por los misioneros de la Provincia de San Francisco de Quito el año 1651*. Imprenta del Ministerio de Gobierno, Quito.

Edmundson, G. 1922. *Journal of the travels and labours of Father Samuel Fritz in the River of the Amazons between 1686 and 1723*. Hakluyt Society, London.

Hemming, J. 1987. *Amazon Frontier: The defeat of the Brazilian indians*. Harvard University Press, Cambridge, Massachusetts.

Jobim, A. 1949. *Monografia geográfica do Estado do Amazonas*. Papelaria Velho Lino, Manaus.

Ramos, A. 1956. *O negro na civilização brasileira*. Livraria-Editôra da Casa do Estudante do Brasil, Rio de Janeiro.

Viola, H. J. 1991. Seeds of change. In *Seeds of change*, edited by H. J. Viola and C. Margolis, pp. 11–15. Smithsonian Institution Press, Washington, D.C.

On landscape change and economic activities in the postcontact period

Agassiz, L. 1896. *A journey in Brazil*. Houghton, Mifflin and Company, Boston.

Anderson, A. B., and E. M. Ioris. 1992. Valuing the rain forest: Economic strategies by small-scale forest extractivists in the Amazon estuary. *Human Ecology* 20.3:337–69.

Anderson, R. L. 1976. *Following curupira: Colonization and migration in Pará, 1758 to 1930 as a study in settlement of the humid tropics*. Ph.D. diss., University of California, Davis.

Anderson, S. D. 1992. Engenhos na várzea: Uma análise do declínio de um sistema de produção tradicional na Amazônia. In *Amazônia: A fronteira agrícola 20 anos depois*, edited by P. Léna and A. E. Oliveira, pp. 101–21. CEJUP, Belém.

Anderson, S. D., M. Nogueira, and F. L. T. Marques. 1993. Tide-generated energy at the Amazon estuary: The use of traditional technology to support modern development. *Renewable Energy* 3(2/3):271–78.

Azevedo, J. L. 1901. *Os Jesuitas no Grão-Pará: Suas missões e a colonização*. Tavares Cardoso & Irmão, Lisbon.

Barham, B. L., and O. T. Coomes. 1994. Reinterpreting the Amazon rubber boom: Investment, the state, and Dutch disease. *Latin American Research Review* 29.2:73–109.

Bates, H. W. 1863. *The naturalist on the river Amazons*. 2 vols. John Murray, London.

Browder, J. O., and B. J. Godfrey. 1997. *Rainforest cities: Urbanization, development, and globalization of the Brazilian Amazon*. Columbia University Press, New York.

Brown, C. B., and W. Lidstone. 1878. *Fifteen thousand miles on the Amazon and its tributaries*. Edward Stanford, London.

Ferreira Penna, D. S. 1869. *A região occidental da Provincia do Pará*. Typographia do Diario, Belém.

Herndon, W. L., and L. Gibbon, 1853. *Exploration of the valley of the Amazon*. Vol. 1. Robert Armstrong, Washington, D.C.

Le Cointe, P. 1903. Le Bas Amazone. *Annales de Géographie* 12:54–66.

———. 1922. *L'Amazonie brésilienne: Le pays-ses habitants, ses resources, notes et statistiques jusqu'en 1920*. 2 vols. Augustin Challamel, Paris.

Lima, R. R. 1956. A agricultura nas várzeas do estuário do Amazonas. *Boletim Técnico do Instituto Agronômico do Norte* 33:1–164.

MacLachlan, C. M. 1974. African slave trade and economic development in Amazônia. In *Slavery and race relations in Latin America*, edited by R. B. Toplin, pp. 112–45. Greenwood Press, Westport, Connecticut.

Marcoy, P. 1869. *Voyage de l'Océan Atlantique à l'Océan Pacifique à travers de l'Amérique du Sud*. L. Hachette, Paris.

Maw, H. L. 1829. *Journal of a passage from the Pacific to the Atlantic, crossing the Andes and descending the River Marañon or Amazon*. John Murray, London.

Oliveira, A. E. 1994. The evidence for the nature of the process of indigenous deculturation and destabilization in the Brazilian Amazon in the last three hundred years. In *Amazon indians: From prehistory to the present*, edited by A. Roosevelt, pp. 95–119. University of Arizona Press, Tucson.

Ramos, A. 1954. *O folclore negro do Brasil*. Livraria Editória da Casa do Estudante do Brasil.

Santa-Anna Nery, F. J. 1910. *The land of the Amazons*. Sands, London.

Santos, R. 1980. *História econômica da Amazônia (1800–1920)*. T. A. Queiroz, São Paulo.

Smith, H. H. 1879. *Brazil: The Amazons and the Coast*. Charles Scribner's Sons, New York.

Spix, J. B., and C. F. P. Martius. 1976. *Viagem pelo Brasil, 1817–1820*. Vol. 3. Edições Melhoramentos, São Paulo.

On confederate settlers in the Amazon

Griffin, T. E. 1981. Confederates on the Amazon. *Américas* 33.2:13–17.

Harter, E. C. 1985. *The lost colony of the Confederacy*. University Press of Mississippi, Jackson.

Hill, L. F. 1927. Confederate exiles to Brazil. *Hispanic American Historical Review* 7:129–210.

Melby, J. F. 1942. The rubber river: An account of the rise and collapse of the Amazon boom. *Hispanic American Historical Review* 22.3:452–69.

Chapter 3

On nontimber forest products

Akers, C. E. 1912. *The Amazon valley: Its rubber industry and other resources*. Waterlow and Sons, London.

Almeida, S. S., P. L. B. Lisboa, and A. S. Silva. 1993. Diversidade florística de uma comunidade arbórea na estação científica "Ferreira Penna," em Caxiuanã (Pará). *Boletim do Museu Paraense Emílio Goeldi, Sér. Bot.* 9.1:93–128.

Anderson, A. B. 1988. Use and management of native forests dominated by açaí palm (*Euterpe oleracea Mart.*) in the Amazon estuary. *Advances in Economic Botany* 6:144–54.

———. 1990. Extraction and forest management by rural inhabitants in the Amazon estuary. In *Alternatives to deforestation: Steps toward sustainable use of the Amazon rain forest*, edited by A. B. Anderson, pp. 65–85, Columbia University Press, New York.

Bahri, S. 1992. *L'Agroforesterie, une alternative pour le développement de la plaine alluviale de l'Amazone: l'Exemple de l'Île de Careiro*. Ph.D. dissertation, Université de Montpellier, Montpellier.

Brabo, M. J. C. 1979. Palmiteiro de Muaná—estudo sobre o processo de produção no beneficiamento do açaizeiro. *Boletim do Museu Paraense Emílio Goeldi*, Nova Série, Antropologia, 73:1–29.

Brondizio, E. S., E. F. Moran, P. Mausel, and Y. Wu. 1994. Land use change in the Amazon estuary: Patterns of caboclo settlement and landscape management. *Human Ecology* 22.3:249–78.

Cavalcante, P. B. 1991. *Frutus comestíveis da Amazônia*. Edições CEJUP, Belém.

Chaves, J. M., and E. Pechnik. 1947. Tucumã. *Revista de Química Industrial* 16.184:5–19.

Clay, J. W. 1996. *Generating income and conserving resources: 20 lessons from the field*. World Wildlife Fund, Washington, D.C.

Ducke, A. 1946. *Plantas de cultura precolombiana na Amazônia brasileira: Notas sôbre as espécies ou formas espontâneas que supostamente lhes teriam dado origem*. Instituto Agronômico do Norte, Boletim Técnico 8, Belém.

Duke, J. A., and R. Vasquez. 1994. *Amazonian ethnobotanical dictionary*. CRC Press, Boca Raton, Florida.
Ferreira, A. R. 1972. *Viagem filosófica pelas Capitanias do Grão Pará, Rio Negro, Mato Grosso, e Cuiabá*. Conselho Federal de Cultura, Rio de Janeiro.
Frikel, P. 1978. Áreas de arboricultura pré-agrícola na Amazônia: Notas preliminares. *Revista de Antropologia* 21.1:45–52.
Henderson, A. 1995. *The palms of the Amazon*. Oxford University Press, New York.
Lange, A. 1914. *The lower Amazon*. Putnam's Sons, New York.
Le Cointe, P. 1947. *Árvores e plantas úteis*. Editora Nacional, Rio de Janeiro.
Padoch, C. 1988. Aguaje (*Mauritia flexuosa* L. f.) in the economy of Iquitos, Peru. *Advances in Economic Botany* 6:214–24.
Pechnik, E., I. V. Mattoso, J. M. Chaves, and P. Borges. 1947. Possibilidade de aplicação do buriti e tucumã na indústria alimentar. *Arquivos Brasileiros de Nutrição* 4.1:33–37.
Pennington, T. D. 1997. *The genus Inga*. Royal Botanic Gardens, Kew.
Plotkin, M. J., and M. J. Balick. 1984. Medicinal uses of South American palms. *Journal of Ethnopharmacology* 10:157–79.
Pollak, H., M. Mattos, and C. Uhl. 1995. A profile of palm heart extraction in the Amazon estuary. *Human Ecology* 23.3:357–85.
Santa-Anna Nery, F. J. 1885. *Le pays des Amazones, l'El-Dorado, les terres a caoutchouc*. L. Frinzine, Paris.
Sauer, J. D. 1979. Living fences in Costa Rican agriculture. *Turrialba* 29.4:255–61.
Strudwick, J., and G. L. Sobel. 1988. Uses of *Euterpe oleracea* Mart. in the Amazon estuary, Brazil. *Advances in Economic Botany* 6:225–53.
Vasquez, R., and A. H. Gentry. 1989. Use and misuse of forest-harvested fruits in the Iquitos area. *Conservation Biology* 3.4:350–61.
Warren, L. A. 1992. *Euterpe palms in northern Brazil: Market structure and socioeconomic implications for sustainable management*. Master's thesis, University of Florida, Gainesville.

On animals as dispersal agents of economically important wild plants

Kubitzki, K. 1985. Dispersal of forest plants. In *Key environments: Amazonia*, edited by G. T. Prance and T. E. Lovejoy, pp. 192–206. Pergamon Press, Oxford.

On fruits and edible seeds from weedy communities or meadows

Bye, Jr., R. 1993. The role of humans in the diversification of plants in Mexico. In *Biological diversity of Mexico: Origins and distribution*, edited by T. P. Ramamoorthy, R. Bye, Jr., A. Lot, and J. Fa, pp. 707–31. Oxford University Press, New York.
Heiser, Jr., C. B. 1992. *Of plants and people*. University of Oklahoma Press, Norman.
Hudson, Jr., J. D. 1986. Relationships of domesticated and wild *Physalis philadelphica*. In *Solanaceae: Biology and systematics*, edited by W. G. D'Arcy, pp. 416–32. Columbia University Press, New York
Yungjohann, J. C. 1989. *White gold: The diary of a rubber cutter in the Amazon 1906–1916*. Edited by G. T. Prance. Synergetic Press, Oracle, Arizona.

On plants used for catching or feeding turtles

Smith, N. J. H. 1974. Destructive exploitation of the South American river turtle. *Yearbook of the Pacific Coast Geographers* 36:85–102.
———. 1979. Aquatic turtles of Amazonia: An endangered resource. *Biological Conservation* 16.3:165–76.
Vieira, A. 1925. *Cartas do Padre António Vieira*. Vol. 1. Imprensa da Universidade, Coimbra.

On medicinal uses of plants

Gilbert, B. 1995. Economic plants of the Amazon: Their industrial development in defense of the forest. In *Chemistry of the Amazon: Biodiversity, natural products, and*

environmental issues, edited by P. R. Seidl, O. R. Gottlieb, and M. A. C. Kaplan, pp. 19–33. American Chemical Society, Washington, D.C.
Mors, W. B. 1995. Poisons and anti-poisons from the Amazon forest. In *Chemistry of the Amazon: Biodiversity, natural products, and environmental issues*, edited by P. R. Seidl, O. R. Gottlieb, and M. A. C. Kaplan, pp. 79–84. American Chemical Society, Washington, D.C.
Schultes, R. E. 1983. De plantis toxicariis e Mundo Novo tropicale commentationes XXXII: Notes, primarily of field tests and native nomenclature, on biodynamic plants of the northwest Amazon. *Botanical Museum Leaflets* (Harvard) 29.3:251–72.
Van den Berg, M. E., and M. H. L. da Silva. 1986. Plantas medicinais do Amazonas. In *Anais do primeiro simpósio do trópico úmido, 12–17 November, Belém, Pará*. Vol. 2, *Flora e Floresta*, pp. 127–33. EMBRAPA, Brasília.

On the timber trade

Adams, M. 1996. Timber trends: Why are export log prices often so much higher than domestic prices? *Tropical Forest Update* (International Tropical Timber Organization) 6.3:18–22.
Ayres, J. M. 1993. *As matas de várzea do Mamirauá*. Conselho Nacional de Desenvolvimento Científico e Tecnológico/Sociedade Civil Mamirauá, Brasília.
Browder, J. O. 1989. Lumber production and economic development in the Brazilian Amazon: Regional trends and a case study. *Journal of World Forest Resource Management* 4:1–19.
Rice, R. E., R. E. Gullison, and J. W. Reid. 1997. Can sustainable management save tropical forests? *Scientific American* 276.4:44–49.
Santos, J. 1988. Diagnóstico das serrarias e das fábricas de laminados e compensados do Estado do Amazonas. *Acta Amazonica* 18.1–2:67–82.

On the abundance of kapok trees before the advent of plywood factories

Gates, R. R. 1927. *A botanist in the Amazon Valley: An account of the flora and fauna in the land of floods*. H. F. & G. Witherby, London.

On calls for deforesting floodplains of tropical rivers, including the Amazon

Camargo, F. C. 1958. Report on the Amazon region. In *Problems of humid tropical regions*, pp. 11–24. UNESCO, Paris.
Pendelton, L. H. 1992. Trouble in paradise: Practical obstacles to nontimber forestry in Latin America. In *Sustainable harvest and marketing of rain forest products*, edited by M. Plotkin and L. Famolare, pp. 252–62. Island Press, Washington, D.C.
Woodroffe, J. F., and H. H. Smith, 1915. *The Rubber industry of the Amazon and how its supremacy can be maintained*. John Bale, Sons & Danielsson, London.

Chapter 4

On the role of fire in shaping floodplain meadows

Betendorf, J. F. 1910. Chronica da missão dos padres da Companhia de Jesus no Estado do Maranhão. *Revista do Instituto Historico e Geografico Brazileiro* 72(1):1–682.
Furtado, L. G. 1993. *Pescadores do Rio Amazonas: Um estudo antropológico da pesca ribeirinha numa área Amazônica*. Museu Paraense Emílio Goeldi, Belém.
Wallace, A. R. 1853. *A narrative of travels on the Amazon and Rio Negro, with an account of the native tribes, and observations on the climate, geology, and natural history of the Amazon valley*. Reeve, London.

On the antiquity of cattle ranching in the Amazon

Avé-Lallemant, 1961. *Viagem pelo norte do Brasil no ano 1859*. Instituto Nacional do Livro, Rio de Janeiro.

Bennett, D., and R. S. Hoffmann. 1991. Ranching in the New World. In *Seeds of change*, edited by H. J. Viola and C. Margolis, pp. 90–111. Smithsonian Institution Press, Washington, D.C.

Bittencourt, A. 1925. *Chorografia do Estado do Amazonas*. Typ. Palacio Real, Manaus.

Eden, M. J. 1990. *Ecology and land management in Amazonia*. Bellhaven Press, London.

Ferreira Penna, D. S. 1876. *A ilha de Marajó*. Tip. do Diário do Grão-Pará, Belém.

Hartt, C. F. 1874. Contributions to the geology and physical geography of the lower Amazonas: The Ereré-Monte Alegre district and the table-topped hills. *Bulletin of the Buffalo Society of Natural History* 26:210–35.

Im Thurn, E. F. 1883. *Among the indians of Guiana: Being sketches chiefly anthropologic from the interior of British Guiana*. Kegan Paul, Trench and Co., London.

Le Cointe, P. 1918. *A Industria pastoril na Amazonia, particularmente no Baixo-Amazonas*. Imprensa Official do Estado, Belém.

Lima, R. R. 1994. *Várzeas da Amazônia brasileira: Principais características e possibilidades*. Faculdade de Ciências Agrárias do Pará, Belém.

Mathews, E. D. 1879. *Up the Amazon and Madeira Rivers, through Bolivia and Peru*. Sampson, Low, Marston, Searle and Rivington, London.

Rivière, P. 1972. *The forgotten frontier: Ranchers of northern Brazil*. Holt, Rinehart and Winston, New York.

Serrão, E. A. 1986. Pastagens nativas do trópico úmido brasileiro: Conhecimentos atuais. In *Anais do Primeiro Simpósio do Trópico Úmido, 12 a 17 de novembro de 1984*. Vol. 5, pp. 83–205. EMBRAPA, Brasília.

On the driving forces behind cattle ranching

Araújo, H. C. 1922. *A Prophylaxia Rural no Estado do Pará*. Typ. da Livraria Gillet, Belém.

Faminow, M. D. 1997. Spatial economics of local demand for cattle products in Amazon development. *Agriculture, Ecosystems, & Environment* 62:1–11.

Furley, P. A., and L. Mougeot. 1994. Perspectives. In *The forest frontier: Settlement and change in Brazilian Roraima*, edited by P. A. Furley, pp. 1–38. Routledge, London.

Hecht, S. B. 1992. Logics of livestock and deforestation: The case of Amazonia. In *Development or destruction: The conversion of tropical forest to pasture in Latin America*, edited by T. E. Downing, S. B. Hecht, H. A. Pearson, and C. Garcia-Downing, pp. 7–25. Westview, Boulder, Colorado.

Hemming, J. 1994. Indians, cattle, and settlers: Growth of Roraima. In *The Forest frontier: Settlement and change in Brazilian Roraima*, edited by P. A. Furley, pp. 39–67. Routledge, London.

Maule, J. P. 1990. *The cattle of the tropics*. Centre for Tropical Veterinary Medicine, University of Edinburgh, Edinburgh.

Mougeot, L., and P. Léna. 1994. Forest clearance and agricultural strategies in northern Roraima. In *The forest frontier: Settlement and change in Brazilian Roraima*, edited by P. A. Furley, pp. 111–52. Routledge, London.

Seré, C., and L. S. Jarvis. 1992. Livestock economy and forest destruction. In *Development or destruction: The conversion of tropical forest to pasture in Latin America*, edited by Theodore E. Downing, Susanna B. Hecht, Henry A. Pearson, and Carmen Garcia-Downing, pp. 95–113. Westview, Boulder, Colorado.

Smith, N. J. H., E. A. S. Serrão, P. Alvim, and I. C. Falesi. 1995. *Amazonia: Resiliency and dynamism of the land and its people*. United Nations University Press, Tokyo.

On the productivity of floodplain grasses

Piedade, M. T. F., W. J. Junk, and S. P. Long. 1991. The productivity of the C4 grass *Echinochloa polystachya* on the Amazon floodplain. *Ecology* 72.4:1456–63.

Piedade, M. T. F., W. J. Junk, and J. A. N. de Mello. 1992. A floodplain grassland of the central Amazon. In *Primary productivity of grass ecosystems of the tropics and sub-*

tropics, edited by S. P. Long, M. B. Jones, and M. J. Roberts, pp. 127–58. Chapman and Hall, London.

Piedade, M. T. F., H. Acquay, M. Biltonen, P. Rice, M. Silva, J. Nelson, V. Lipner, S. Giordano, A. Horowitz, and M. D'Amore. 1992. Environmental and economic costs of pesticide use: An assessment based on currently available U.S. data, although incomplete, tallies $8 billion in annual costs. *Bioscience* 42:750–60.

On the introduction of African grasses to the New World

Parsons, J. 1972. Spread of African pasture grasses to the American tropics. *Journal of Range Management* 25.1:12–17.

On nutritional disorders of cattle

Sutmöller, P., V. Abreu, J. van der Grift, and W. G. Sombroek. 1966. *Mineral imbalances in cattle in the Amazon Valley*. Communication 53, Koninklijk Instituut voor de Tropen, Amsterdam.

On the introduction and spread of water buffalo

Alvim, P. 1990. Agricultura apropriada para o uso contínuo dos solos na região Amazônica. *Espaço, Ambiente e Planejamento* 2.11:3–71.

Chibnik, M. 1994. *Risky rivers: The economics and politics of floodplain farming in Amazonia*. University of Arizona Press, Tucson.

Fonseca, W. 1987. *Búfalo: Estudo e comportamento*. Icone Editora, São Paulo.

Leal, C. 1990. No Marajó, a beleza confunde com velhos problemas. *O Liberal* (Belém), 29 May, p. 18.

Lima, R. R., and M. M. Tourinho. 1994. *Várzeas da costa Amapaense: Características e possibilidades agropecuárias*. Faculdade de Ciências Agrárias do Pará, Belém.

Mahadevan, P. 1974. The buffaloes of Latin America. In *The husbandry and health of the domestic buffalo*, edited by W. Cockrill, pp. 676–704. Food and Agriculture Organization, Rome.

Ohly, J. J. 1986. Water-buffalo husbandry in the central Amazon region. *Animal Research and Development* 24:23–40.

Ohly, J. J., and M. Hund. 1996. Pasture farming on the floodplains of central Amazonia. *Animal Research and Development* 43/44:53–79.

On the ecological and cultural impacts of ranching

McGrath, D. G., F. Castro, C. Futemma, B. D. Amaral, and J. Calabria. 1993. Fisheries and the evolution of resource management on the lower Amazon floodplain. *Human Ecology* 21.2:167–95.

On goat and pig raising

Anderson, A. B., A. Gely, J. Strudwick, G. L. Sobel, and M. G. C. Pinto. 1985. Um sistema agroflorestal na várzea do estuário Amazônico (Ilha das Onças, Município de Bacarena, Estado do Pará). *Acta Amazonica*, Supl. 15.1–2:195–224.

Smyth, W., and F. Lowe. 1836. *Narrative of a journey from Lima to Pará, across the Andes and down the Amazon*. John Murray, London.

On buriti fruits as source of food for game

Bodmer, R. E. 1989. Frugivory in Amazonian artiodactyla: Evidence for the evolution of the ruminant stomach. *Journal of the Zoological Society of London* 219:457–67.

———. 1991. Strategies of seed dispersal and seed predation in Amazonian ungulates. *Biotropica* 23.3:255–61.

On capybara raising

Dourojeanni, M. J. 1985. Over-exploited and under-used animals in the Amazon region. In *Key environments: Amazonia*, edited by G. T. Prance and T. E. Lovejoy, pp. 192–206. Pergamon Press, Oxford.

Chapter 5

On the origins, cultivation, and processing of jute

Benchimol, S. 1992. *Amazônia: A Guerra na Floresta*. Civilização Brasileira, Rio de Janeiro.

———. 1992. Amazônia interior: Apologia e holocausto. In *Amazônia: Desenvolvimento ou Retrocesso*, edited by J. M. Costa, pp. 231–63. Edições CEJUP, Belém.

Bergman, R. W. 1980. *Amazon economics: The simplicity of Shipibo indian wealth*. Dellplain Latin American Studies 6, Department of Geography, Syracuse University, Syracuse, New York.

Biard, J., and G. Wagenaar. 1960. Crop production in selected areas of the Amazon Valley. *FAO ETAP Report* 1254:1–47.

Dempsey, J. M. 1975. *Fiber crops*. University Presses of Florida, Gainesville.

Homma, A. K. O. 1995. A civilização da juta na Amazônia: Expansão e declínio. In *Anais do congresso brasileiro de economia e sociologia rural*. Vol. 1, pp. 509–31. SOBER, Brasília.

Libonati, V. F., and L. P. Soares. 1966. Problemas atuais da juticultura Amazônica. *Pesquisa Agropecuária Brasileira* 1:1–6.

Mors, W. B., and Carlos T. Rizzini. 1966. *Useful plants of Brazil*. Holden-Day, San Francisco.

Slater, C. 1994. *The dance of the dolphin: Transformation and disenchantment in the Amazonian imagination*. University of Chicago Press, Chicago.

Thigpen, M. E., P. Marongiu, and S. R. Lasker. 1987. *World demand prospects for jute*. World Bank Staff Commodity Working Papers, no. 16, Washington, D.C.

Wesche, R., and T. Bruneau. 1990. *Integration and change in Brazil's middle Amazon*. University of Ottawa Press, Ottawa.

On pesticides and pest control

Murrieta, R. S. S., E. Brodízio, A. Siqueira, and E. F. Moran. 1992. Estratégias de subsistência da comunidade de Praia Grande, Ilha de Marajó, Pará, Brasil. *Boletim do Museu Paraense Emílio Goeldi*, Série Antropologia 8.2:185–201.

On traditional maize varieties

Brigham, W. T. 1887. *Guatemala, the land of the Quetzal*. Charles Scribner's Sons, New York.

On the history of rice cultivation and yields

Baena, A. L. M. 1969. *Compêndio das eras da Província do Pará*. Universidade Federal do Pará, Belém.

Barrow, C. J., and A. Paterson. 1994. Agricultural diversification: The contribution of rice and horticultural producers. In *The forest frontier: Settlement and change in Brazilian Roraima*, edited by P. A. Furley, pp. 153–84. Routledge, London.

Camargo, F. C. 1948. Terra e colonização no antigo e novo Quaternário da Zona da Estrada de Ferro de Bragança, Estado do Pará, Brasil. *Boletim do Museu Goeldi* 10:123–33.

Fearnside, P. M. 1992. Desmatamento e desenvolvimento agrícola na Amazônia brasileira. In *Amazônia: A Fronteira Agrícola 20 Anos Depois*, edited by P. Léna and A. E. Oliveira, pp. 207–22. CEJUP, Belém.

Katzman, M. T. 1975. Regional development policy in Brazil: The role of growth poles and development highways in Goiás. *Economic Development and Cultural Change* 24:75–107.

Kinkead, G. 1981. Trouble in D. K. Ludwig's jungle. *Fortune* 103.8:102–15.

On manioc origins, varieties, and yields

Albuquerque, M. 1969. *A mandioca na Amazônia*. SUDAM, Belém.

———. 1973. *Cultura da Mandioca*. Circular 16, PEAN/ACAR, Belém.

Boster, J. 1983. A comparison of the diversity of Jivaroan gardens with that of the tropical forest. *Human Ecology* 11.1:47–68.

———. 1984. Classification, cultivation, and selection of Aguarana cultivars of Manihot esculenta (*Euphorbiaceae*). *Advances in Economic Botany* 1:34–47.

Carneiro, R. L. 1983. The cultivation of manioc among the Kuikuru of the Upper Xingu. In *Adaptive responses of native Amazonians*, edited by R. B. Hames and W. T. Vickers, pp. 65–111. Academic Press, New York.

Chernela, J. M. 1986. Os cultivares de mandioca na área do Uaupés (Tukâno). In *Suma etnológica brasileira: 1. Etnobiologia*, edited by B. G. Ribeiro, pp. 151–58. Vozes, Petrópolis.

Plotkin, M. J. 1993. *Tales of a shaman's apprentice: An ethnobotanist searches for new medicines in the Amazon rain forest*. Viking, New York.

Salick, J., N. Cellinese, and S. Knapp. 1997. Indigenous diversity of cassava: Generation, maintenance, use and loss among the Amuesha, Peruvian upper Amazon. *Economic Botany* 51.1:6–19.

Schultes, R. E. 1979. The Amazonia as a source of new economic plants. *Economic Botany* 33.3:259–66.

On control of leaf-cutter ants in manioc fields

Sousa, G. S. 1971. *Tratado descritivo do Brasil em 1587*. Editora Nacional/Editôra da Universidade de São Paulo, São Paulo.

On squash varieties

Heiser, Jr., C. B. 1992. *Of plants and people*. University of Oklahoma Press, Norman.

Tindall, H. D. 1983. *Vegetables in the tropics*. Macmillan, London.

Chapter 6

On the suitability of perennial crops in flood-prone areas

Junk, W. J. 1989. Flood tolerance and tree distribution in central Amazonian floodplains. In *Tropical forests: Botanical dynamics, speciation and diversity*, edited by L. B. Holm-Nielsen, I. C. Nielsen, and H. Balslev, pp. 47–64. Academic Press, London.

Leeuwen, J., F. C. T. Costa, F. A. Catique, M. M. Pereira, M. Woude, C. L. Hemmes, J. B. Gomes, P. Viana. 1993. Agroforestry technology development with farmers in central Amazonia: Research in progress. In *Management and rehabilitation of degraded lands and secondary forests in Amazonia*, edited by J. A. Parrotta and M. Kanashiro, pp. 188–90. International Institute of Tropical Forestry, USDA-Forest Service, Rio Pedras, Puerto Rico.

On agroforestry on the Amazon floodplain

Guillaument, J., M. Lourd, S. Bahri, and A. A. Santos. 1993. Os sistemas agrícolas na Ilha do Careiro. *Amazoniana* 12.3/4:527–50.

Leeuwen, J. 1992. Towards establishing agroforestry research priorities for central Amazonia. In *Annals Forest 90: First international symposium on environmental studies on tropical rain forests, 7–13 October, Manaus, Brazil*, pp. 113–17. Sociedade Brasileira para a Valorização do Meio Ambiente (BIOSFERA), Rio de Janeiro.

Smith, N. J. H., I. C. Falesi, P. T. Alvim, and E. A. S. Serrão. 1996. Agroforestry trajectories among smallholders in the Brazilian Amazon: Innovation and resiliency in pioneer and older settled areas. *Ecological Economics* 18:15–27.

Smith, N. J. H., T. J. Fik, P. Alvim, I. C. Falesi, and E. A. S. Serrão. 1995. Agroforestry developments and potential in the Brazilian Amazon. *Land Degradation and Rehabilitation* 6:251–63.

On home gardens on the Amazon floodplain

Hiraoka, M. 1985. Floodplain farming in the Peruvian Amazon. *Geographical Review of Japan* 58 (ser. B, no. 1):1–23.

Smith, N. J. H. 1996. Home gardens as a springboard for agroforestry development in Amazonia. *International Tree Crops Journal* 9:11–30.

Sternberg, H. O'R. 1995. Waters and wetlands of Brazilian Amazonia: An uncertain future. In *The fragile tropics of Latin America: Sustainable management of changing environments*, edited by T. Nishizawa and J. I. Uitto, pp. 113–79. United Nations University Press, Tokyo.

On the diversity of banana cultivars

Stover, R. H., and N. W. Simmonds. 1987. *Bananas*. Longman Scientific & Technical, Burnt Mill.

On the early stages of plant domestication

Harris, D. R. 1978. The agricultural foundations of lowland Maya civilization: A critique. In *Pre-Hispanic Maya agriculture*, edited by P. D. Harrison and B. L. Turner, pp. 301–22. University of New Mexico Press, Albuquerque.

Rindos, D. 1984. *The origins of agriculture: An evolutionary perspective*. Academic Press, Orlando.

On fruit in the diet of tambaqui

Araújo-Lima, C., and M. Goulding. 1997. *So fruitful a fish: Ecology, conservation, and aquaculture of the Amazon's tambaqui*. Columbia University Press, New York.

Ducke, A. 1925. Plantes nouvelles ou peu connues de la région amazonienne (IIIe partie). *Archivos do Jardim Botânico do Rio de Janeiro* 4:1–205.

On logging and virola stocks

Anderson, A. B., I. Mousasticoshvily, Jr., and D. S. Macedo. 1994. *Impactos Ecológicos e Sócio-Econômicos da Exploração Seletiva de Virola no Estuário Amazônico*. World Wildlife Fund, Brasília.

Chapter 7

On the suitability of Amazonia for sustainable agriculture

Meggers, B. J. 1992. Prehistoric population density in the Amazon basin. In *Disease and demography in the Americas*, edited by J. W. Verano and D. H. Ubelaker, pp. 197–205. Smithsonian Institution Press, Washington, D.C.

———. 1995. Archaeological perspectives on the potential of Amazonia for intensive exploitation. In *The fragile tropics of Latin America: Sustainable management of changing environments*, edited by T. Nishizawa and J. I. Uitto, pp. 68–93. United Nations University Press, Tokyo.

———. 1995. Judging the future by the past: The impact of environmental instability on prehistoric Amazonian populations. In *Indigenous peoples and the future of Amazonia: An ecological anthropology of an endangered world*, edited by L. E. Sponsel, pp. 15–43. University of Arizona Press, Tucson.

On community resource management

Alexander, B. 1994. People of the Amazon fight to save the flooded forest. *Science* 265: 606–7.

Alvarado, D. F. 1994. Programa integral de desarrollo y conservación Pacaya-Samiria. In *Trópico em movimento: Alternativas contra a pobreza e a destruição ambiental no trópico úmido*, pp. 163–81. POEMA, Belém.

Ayres, D. L., and J. M. Ayres. 1995. Mamirauá: Ribeirinhos e a preservação da biodiversidade da várzea amazônica. In *Abordagens interdisciplinares para a conservação da biodiversidade e dinâmica do uso da terra no Novo Mundo*, edited by G. A. B. Fonseca, M. Schmink, L. P. Pinto, and F. Brito, pp. 169–82. Conservation International do Brasil, Universidade Federal de Minas Gerais, University of Florida, Belo Horizonte.

Ayres, J. M., E. A. F. Moura, D. Lima-Ayres. 1994. Estação Ecológica Mamirauá: O desafio de preservar várzea na Amazônia. In *Trópico em movimento: Alternativas contra a pobreza e a destruição ambiental no trópico úmido*, pp. 35–52. POEMA, Belém.

Bodmer, R. E., and J. M. Ayres. 1991. Sustainable development and species diversity in Amazonian forests. *Species* 16:22–24.

Léna, P., C. Geffray, and R. Araújo. 1996. Avant-propos: L'oppression paternaliste au Brésil. In *L'Oppression paternaliste au Brésil*, pp. 105–8. Éditions Karthala, Paris.

On cultural attributes of crop varieties

Nazarea, V. D., E. Tison, M. C. Piniero, R. E. Rhoades. 1997. *Yesterday's ways, tomorrow's treasures: Heirloom plants and memory banking*. Kendall/Hunt, Dubuque, Iowa.

On resiliency and openness in agricultural systems

Hiraoka, M. 1980. *Japanese agricultural settlement in the Bolivian upper Amazon: A study in regional ecology*. Latin American Studies, no. 1, Special Research Project on Latin America, University of Tsukuba, Sakuramura, Ibaraki, Japan.

Wilbanks, T. J. 1994. "Sustainable development" in geographic perspective. *Annals of the Association of American Geographers* 84:541–56.

Index